SpringerBriefs in Materials

The SpringerBriefs Series in Materials presents highly relevant, concise monographs on a wide range of topics covering fundamental advances and new applications in the field. Areas of interest include topical information on innovative, structural and functional materials and composites as well as fundamental principles, physical properties, materials theory and design.SpringerBriefs present succinct summaries of cutting-edge research and practical applications across a wide spectrum of fields. Featuring compact volumes of 50 to 125 pages, the series covers a range of content from professional to academic. Typical topics might include

- A timely report of state-of-the art analytical techniques
- A bridge between new research results, as published in journal articles, and a contextual literature review
- A snapshot of a hot or emerging topic
- An in-depth case study or clinical example
- A presentation of core concepts that students must understand in order to make independent contributions

Briefs are characterized by fast, global electronic dissemination, standard publishing contracts, standardized manuscript preparation and formatting guidelines, and expedited production schedules.

More information about this series at http://www.springernature.com/series/10111

Adam A. Tracy • Sujata K. Bhatia
Krish W. Ramadurai

Bio-Based Materials as Applicable, Accessible, and Affordable Healthcare Solutions

 Springer

Adam A. Tracy
Harvard University
Cambridge, MA, USA

Krish W. Ramadurai
Harvard University
Cambridge, MA, USA

Sujata K. Bhatia
Chemical & Biomolecular Engineering
University of Delaware
Newark, DE, USA

ISSN 2192-1091   ISSN 2192-1105   (electronic)
SpringerBriefs in Materials
ISBN 978-3-319-69325-5   ISBN 978-3-319-69326-2   (eBook)
https://doi.org/10.1007/978-3-319-69326-2

Library of Congress Control Number: 2017959748

Printed on acid-free paper

This Springer imprint is published by Springer Nature
The registered company is Springer International Publishing AG
The registered company address is: Gewerbestrasse 11, 6330 Cham, Switzerland

To the patients and healthcare workers striving for better outcomes. And to the innovators of the future, those from the past send greetings.

Preface

Just as our world finds itself in a perpetual state of change and growth, the healthcare field is also immersed in a dynamic environment with ever-changing problems and solutions. In responding to these challenges, an extensive amount of effort has focused on reforming healthcare systems, implementing disruptive technologies, and fostering innovative solutions. While this integrative approach is applaudable, these interventions are often times framed within a microscopic lens, focusing on only a local scale. As our world becomes increasingly globalized, the ability to foster these initiatives in healthcare on a global, macroscopic scale becomes imminent. Healthcare in developing countries has noticeably improved over the past century and has emerged into a booming economic sector; however, there is great potential and opportunity for growth.

Medical devices and technologies in our current day and age have seen rapid advancements, both in scope of application and interventional capacity, leading to vastly improved patient outcomes and survival. While this is indeed a great achievement, the development, deployment, and effective implementation of these resources often times is not suitable for developing countries which lack adequate infrastructure to support these resources and require more basic ways to provide treatment. Moore's law, a paradigm that considers advancement synonymous with increased digitization and optimization of electronic processes, defines the history of technology. However, the functionality of advanced and "smart" technology is often times futile and impractical in areas that are underdeveloped and do not have the infrastructure to support them. These regions lack some of the most basic elements for medical technologies to properly function and deliver services that are vital for improving human health. Such elements that are often limited or do not exist include electrical infrastructure, accessibility to accessory components, computer analysis tools, and network architecture. These elements, coupled with poor physical infrastructure and lack of human capital to support technological implementation, are real functional barriers for entry and sustainability of these technologies. Rather than importing medical devices from industrialized countries, we propose that the mindset and research focus for developing countries must be on "successful" technologies. In particular, we as researchers seek to demonstrate that effective medical devices deployed in developing countries do not require complex infrastructure and extensive protocols but rather frugally engineered innovations and technology. These frugal innovations which are the most basic, broken-down component technologies can be adapted for an array of environments and countries and serve to break down barriers to medical access and delivery. Simply put, developing countries need technology and innovations that "get the job done." A time has come to create a paradigm shift in developing countries where they can harness the power of intellectual discovery to create novel, yet feasible, interventions to improve healthcare systems in their respective countries. The global burden of disease will continue to further perpetuate, and the ability of countries to harness the ability to develop their own unique frugal technologies to empower the health and wealth of their people will be imminent.

In this book, we explore the country of Nigeria, the most populous country in Africa and a region in the lowest income group per capita. We use Nigeria to demonstrate the potential for healthcare innovation, reorganization, and collaboration with the introduction of "successful" and frugally engineered technologies centered around the deployment of bio-based and sustainable natural resources for use in medical device applications. Bio-based materials derived from agriculture such as corn and soy are currently engineered and utilized in advanced countries to treat an array of medical conditions and maladies. Functional utility lies in the ability to adapt these scientific advancements to be applicable to improve outcomes for the most pressing medical needs in Nigeria while simultaneously improving the economic and healthcare systems. It is in the eyes of innovation that we see the future of our world.

Boston, MA, USA   Adam A. Tracy
Newark, DE, USA   Sujata K. Bhatia
Lemont, IL, USA   Krish W. Ramadurai

Acknowledgments

I owe my deepest gratitude to my mentor, Dr. Sujata K. Bhatia, for her kind support and outstanding guidance. This has been a challenging yet rewarding experience and it was her dedication, positive energy, and global vision that inspired me to address this complicated healthcare need. My family, friends, and colleagues deserve generous thanks for their endless encouragement and understanding. We would also like to acknowledge and thank our coauthor, Krish Ramadurai at Harvard University's Belfer Center for Science and International Affairs, for his contributions in further exploring the unique innovations in this book.

Acknowledgments

Contents

List of Figures

List of Tables

Chapter 1
Introduction

In 1979, the World Health Organization released the report, *Formulating Strategies for Health for All by the Year 2000*, that sought to address the global burden of disease with monetary investments from government agencies, industries, and charities as well as the donation of medical devices (WHO 1979). While indeed the intentions of this project were pure and meant to combat the global burden of disease, it was ultimately wrought with failure due to a myriad of unintended elements. The failure of this project was primarily due to the mistrust in government by local communities, expensive installation and maintenance of engineered solutions, and high levels of emigration of health professionals, all of which disincentivize large public projects (WHO 2010a). This effort, while noble, proved to be unsustainable in its interventional scope and capacity as the functional burden of disease has continued to grow and the resources allocated have either been depleted or deemed impractical. Developing countries face unique challenges in establishing and sustaining adequate primary care to a majority of their citizens, and the role of technology in serving as an impetus for improving and determining healthcare outcomes is more pronounced than ever.

A recent report by Dada et al. (2013) highlights Nigeria's abundant natural resources and chemical engineering talent. A major shift is being observed where natural gas sectors are projected to boost the economy and restore growth in the chemical process industries (Dada et al. 2013). With major reforms and policy adjustments in the power sector being issued by the federal government of Nigeria, there is expected to be substantial growth in domestic manufacturing (Dada et al. 2013). In addition, there is increased investment in agricultural production for use in Africa's industrial sectors (Africa Progress Panel 2014). Realization of these combined measures could have a significant impact on the Nigerian healthcare sector. One may ponder what the connection between domestic manufacturing/ engineering operations and the healthcare sector is, and this is certainly a valid question. The connection lies in the ability to domestically harbor natural, bio-based materials, which can be developed, engineered, and delivered on a local scale to

© The Author(s) 2018
A.A. Tracy et al., *Bio-Based Materials as Applicable, Accessible, and Affordable Healthcare Solutions*, SpringerBriefs in Materials,
https://doi.org/10.1007/978-3-319-69326-2_1

serve people in that immediate region. While the biomaterials market is vast and growing globally, its function and impact in developing countries is minimal (Young 2003). This case study aims to make the connections between Nigeria's agricultural and engineering potential and healthcare needs with advancements in biomaterials science and applied engineering.

Healthcare is a universal necessity and considered a fundamental human right. While we consider healthcare as vital to the health and wealth in our societies, there is a large dissonance in the accessibility and delivery of healthcare services among countries around the world. Countries such as the United States, Australia, England, and Canada and a host of other countries in Europe have access to the most innovative medical technologies at their fingertips. However, a large majority of the global population outside of these countries lacks access to even the most basic, fundamental healthcare needs and supplies. Developing countries such as Nigeria, which have limited healthcare infrastructure, scarce resources, and lack adequate human capital in the form of medical providers, are highly susceptible to the burdens of disease and illness (WHO 2013a). This susceptibility fosters a perpetual cycle of illness in the populace, as individuals that are sick continue to become sick in a cyclical fashion since they cannot receive adequate medical treatment. This not only affects their own health and economic livelihood but also the state of entire communities and ultimately the country itself. A healthy population and labor force is vital for any country but holds true for developing countries especially. Many developing countries harbor agriculturally based economies, in which farmers becoming perpetually sick can have catastrophic economic, social, and fiscal impacts on their families and ultimately the economic output of the country. As the burden of disease continues to threaten developing countries, the need for novel biomaterials and innovative solutions becomes even more vital (Bhatia 2010).

Innovation is the driving force behind creative discovery and exploration in any country; however, we focus on how we can deliver and define these innovations in a feasible context for developing countries such as Nigeria. The founder and CEO of *Design that Matters*, Timothy Prestero, shed light on this topic in his June 2012 TED talk. In his inspirational talk, he focused on the engineering phenomenon of product design for awards and accolades. He insisted that instead of recognition, engineers should adjust their attention by not only designing the product itself but also *designing outcomes* (TEDxBoston 2012). The purpose of this case study is to use this approach to demonstrate that "successful" technologies can serve as powerful, yet discerning driving forces for enhancing social, economic, and public health welfare and alleviating some of the most pressing health issues faced by societies today. In going back to the utilization of bio-based materials in improving healthcare, one may have many questions. What are these materials? Where are the materials sourced? Is this a feasible solution? How do they seek to promote improved health outcomes in countries such as Nigeria? What role do they have in combating the global burden of surgical disease? These questions regarding biomaterials are very valid and are the ones that we seek to answer and further develop throughout the course of this book. One fundamental concept that we center our case on Nigeria is that the versatility of biomaterials can indeed serve as a vital element in defining the future of healthcare services in Nigeria as well as tool that can empower Nigeria in improving health of its people and country as a whole.

The Bio-Economy, Bio-Based Materials, and Frugal Engineering

Bio-based materials derived from renewable agricultural resources are recognized as biocompatible materials that can functionally interact with biomolecules, support the body's natural structures, and perform the desired outcome without causing an adverse host response (Bhatia 2010). These materials are unique in that they are naturally derived, can be locally sourced, are highly biocompatible, and can be utilized in an array of engineering disciplines. In particular these materials, which vary from bio-based building blocks such as polylactic acid (PLA) to natural polymers such as starch and cellulose, can be utilized in a multitude of applications, ranging from bioplastics to biofuels. The integration of these biomaterials into working economies presents a unique opportunity. Chemical Engineering Progress (CEP), a flagship publication of the American Institute of Chemical Engineers (AIChE), published a report by Murali M. Reddy et al. (2012) that highlights a paradigm shift that will occur as our global economies progressively shift from a fossil-fuel- and petroleum-based economy to what is known as a "bio-economy." This new type of economy will be centered around the use of agricultural crops, crop-derived by-products, forestry resources, and other forms of renewable biomass to produce bio-based fuels and bio-based materials, also referred to as biopolymers or bio-derived materials (Fig. 1.1) (Reddy et al. 2012).

These bio-based materials that will play a role in the future of bio-economies are made from substances derived, in whole or in part, from renewable sources and living matter, and can be classified into three main categories based on their origin of production (Weber 2000):

- *Biomass derived.* These bio-based materials are directly extracted or removed from biomass. Some examples include polysaccharides (carbohydrates) such as starch from potatoes and maize; cellulose in wood; carrageenan; pectin; chitin; proteins from soy, corn, and silk; casein; whey; gluten; and cross-linked triglyceride lipids.
- *Biomonomer derived.* These bio-based materials are produced by traditional chemical synthesis using monomers acquired from renewable agricultural resources. These can include polyesters, polylactates, and more specifically the widely used polylactic acid (PLA), a biopolymer that is synthesized from lactic acid monomers that may be produced through fermentation of agricultural carbohydrate feedstocks, such as corn starch (Fig. 1.2).
- *Microorganism derived.* These bio-based materials come from microorganisms or genetically modified bacteria. The polymers produced by these organisms consist mainly of polyhydroxyalkanoates (PHA); however, progress is being

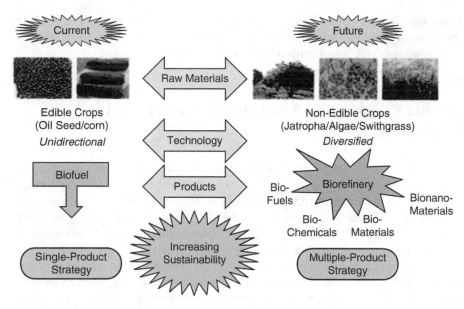

Fig. 1.1 Schematic of the future bio-economy, where greater diversity in raw materials can lead to a wider range of products and increased sustainability (Reddy et al. 2012)

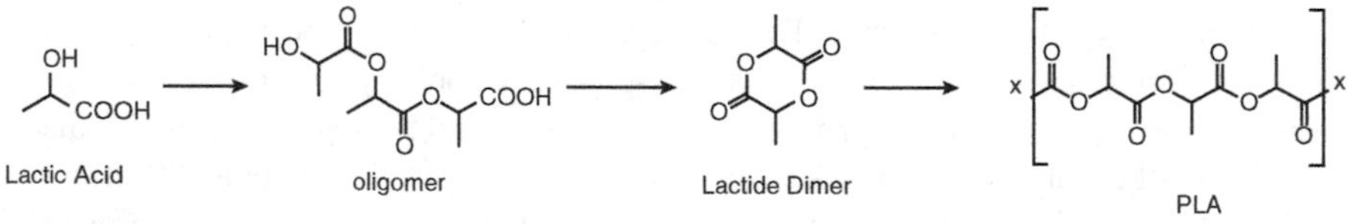

Fig. 1.2 Chemical synthesis of polylactic acid (PLA) from lactic acid monomers (USDA 2008)

made with bacterial cellulose, xanthan, and curdlan, a high molecular weight polymer of glucose.

It is important to note that only the starch-based biomaterials, depicted in the first category, can be produced without advanced biotechnology. The other biomaterials described require exhaustive fermentation techniques and microbiological methods in order to adequately synthesize them for use.

In introducing the idea of a "bio-economy," we must decipher the use and application of this economy for the benefit of humanity. All economies serve a distinct, functional purpose, and in further examining the relative tenets and functional basis for a "bio-economy," one might argue that it does not constitute a novel idea or entity. The reality is that a transition to a bio-economy will address issues on multiple fronts including those related to food security, health, industrial restructuring, and energy security (Bugge et al. 2016). The difficulty lies in deciphering the shear breadth of what constitutes a bio-economy. Researchers have determined that there are three distinct versions or "visions" related to the development of a bio-economy. First, the "biotechnology" version or vision emphasizes the importance of biotechnology research and application and commercialization of biotechnology in different sectors of the economy (Bugge et al. 2016). The second version focuses on "bio-resources," specifically that of processing and upgrading of biological raw materials (Bugge et al. 2016). The third and final version focuses on "bioecology," including the sustainability and ecological processes that optimize the use of energy and promote biodiversity (Bugge et al. 2016). For the sake of this book, we will recognize the bio-economy as an integrative and nonsegregated entity that surmises these three versions. However, we will focus upon the role of specific elements related to each version of the bio-economy and how it will relate to specific innovations.

Breakthroughs in multidisciplinary research over the past two decades have led to commercial success for bio-based materials derived from annually renewable resources (Young 2003). Although still in its infancy, the industry of bio-based products has potential to significantly impact a broad range of industries. One specific sector, bio-based polymers, has demonstrated promise as materials are comparable to traditional plastics in both cost and performance (Reddy 2012). The bioplastic industry is growing at an annual rate of 30%, compared to a 5% growth rate of traditional plastics. Additionally, market research indicates the global production of bioplastics will reach 2.33 million tons this year (Shen et al. 2009). Currently, bio-based plastics only account for 1% of the total plastic production worldwide; however, the USDA predicts that bio-based plastics could replace up to one third of traditional plastics (Shen et al. 2009; USDA 2008). The potential for the bio-based sector is understood through sustained research, and it is evident that continued development, commercialization, adoption, and implementation into the world of healthcare will have a significant impact on the availability and sustainability of medical products and devices.

As the global burden of disease continues to disproportionally afflict people in developing countries, the need for novel biomaterials that can be utilized in the fabrication of novel, low-cost medical devices becomes apparent. Fabrication of novel bio-based medical devices could serve as valuable alternatives to traditional medical devices/supplies that are often expensive and inadequately suited to be deployed in developing countries. In addition to these bio-based medical devices being manufactured and distributed locally, they also harbor unique biomaterials properties that make them highly suited for deployment in medical settings. Bio-based polymers are generated from naturally derived and renewable agricultural resources and are increasingly recognized as biocompatible materials that can functionally interact with and support the body's natural structures (Bhatia 2010). This complementary interaction with human tissues, blood, and natural structures allows the devices to be biologically inert and still perform their respective functions. To illustrate a few examples, bio-based materials, such as polylactic/polyglycolic acid and soybean extracts, have demonstrated success in wound closure and bone repair (Bhatia et al. 2007; Merolli et al. 2010). Numerous commercially available biodegrading sutures are woven from polymers found in corn. Soybean-based materials are being used to generate bone fillers for vertebral fractures and fixation of implants; and in addition, phospholipids derived from soybeans are used to form liposomes for efficient encapsulation and targeted delivery of several drugs (Santin and Ambrosio 2008; Bakker-Woudenberg et al. 1993). Bio-based materials could indeed serve to create a new type of economic entity, promote environmentally responsible practices, and, most importantly, enhance the interventional capacity of healthcare institutions and maximize patient outcomes in developing countries such as Nigeria.

While we have discussed the integration of biomaterials development and the basic elements of a bio-economy, in combating the burden of disease in countries such as Nigeria, the need for "disruptive innovation" and "frugal engineering" becomes imperative. Disruptive innovations refer to technologies that significantly impact the way a particular market functions (Christensen and Raynor 2013). These technologies are not a reiteration of a previous technology, but rather are completely novel in nature and "disrupt" the market space and the previous technologies occupying it. Disruptive innovations not only include disruptive technologies but also a novel application or use of a technology (Christensen and Raynor 2013). These innovations, such as the ones that we explore in this book, can be distinctly geared toward preserving and enhancing the quality of life for others. Disruptive innovations have the unique functional capacity to drive rapid improvements in the healthcare systems as well as improve the social, economic, and individual health outcomes of populations. Disruptive innovations can serve as an impetus for change and effectively foster a paradigm shift in which integrative, yet targeted solutions can be derived to address the most pertinent health threats facing developing countries such as Nigeria today. Bio-based medical devices and supplies can disrupt the current market space of traditional medical devices and supplies that are often costly and difficult to allocate in resource-poor settings (Bhatia and Ramadurai 2017). In addition to disruptive innovations, the concept of frugal engineering serves as an equally important component that can drive future applications of advanced technologies in resource-poor settings.

Frugal engineering is essentially a minimalist approach to innovation and is defined as the process of reducing the complexity and cost of a good and its associated production (Maric et al. 2016). The core concept of frugal engineering involves taking a technology that has been manufactured for use and deployment in advanced infrastructure settings such as high-income countries and essentially breaking it down into its fundamental components to meet the needs of consumers in resource-constrained settings such as those of LMICs (Maric et al. 2016). In deploying the principles of frugal engineering, technologies that may have not been able to be feasibly deployed or implemented in resource-poor settings can indeed be utilized and further adapted to meet the needs of its respective environment. In addition to physically breaking down technologies, frugal engineering seeks to redefine other aspects related to a technology's development and relative application. In exploring the concept of frugal engineering, the components of frugality plus innovation are integrated, defining what is known as "frugal innovation" (Bhatti 2012).

Frugal innovation is the culmination of frugal engineering and disruptive innovation and is a vital element in the design of future bio-based medical devices and innovations. Frugal innovation related to medical devices derived from biomaterials can redefine the physical design, manufacturing process, and distribution of medical devices/supplies in Nigeria. In modifying these various elements related to a technology, one can radically alter the application and adaptability complex of various technologies to be suited for markets in resource-constrained settings. This includes primarily reducing the material cost of these devices with the use of novel biomaterials so that they can be both affordable and accessible. Coupling the fundamental concepts of disruptive innovation with that of the fundamental elements of frugal engineering can open the door to unparalleled possibilities when creating novel bio-based medical device applications in countries such as Nigeria.

Biomedical Material Specifications

Before discussing the unique potential of bio-based materials to influence human health and treatment outcomes, it is vital to define the scope of biomaterials science and describe the ideal biomedical material (Bhatia 2010). Just over two decades ago, a group of biomaterials experts established what is now a commonly used definition for a biomaterial, which is "a nonviable material used in a medical device, intended to interact with biological systems" (Williams 1987). To clarify the use of this definition for this report, the material itself is nonviable; however, it may be extracted, synthesized, and modified from a viable source, such as agricultural products. The most important and defining characteristic of a biomedical material is that of "biocompatibility." This refers to the ability of the material to function appropriately in the human body and to perform the desired clinical result without causing an adverse host response. The biomedical material and any products associated with its degradation in the body must not interfere with wound healing or induce a host immune response. The use of biomaterials for medical purposes requires stringent precautions where physical, biological, chemical, and mechanical characteristics must be similar to the natural physiological environment (Bhatia 2012). These materials must be nontoxic, noninflammatory, efficiently interact with surrounding tissues, and react to biological and environmental stimuli (Williams 1987).

The goal of an ideal biomaterial is to provide effective treatment of a medical malady such as tissue or organ damage. In this case, these materials must have the distinct ability to foster in vivo healing and repair, by complementarily working with the body's natural healing processes. These particular classes of biological elements can supplement and stimulate the body's intrinsic immunologic capacities for tissue regeneration, thus allowing for the bio-based material as well as the structure it comprises, to seamlessly interact with the living tissue (Greenwood et al. 2006). In addition, bio-stability and anatomical site deployment serve as critical confounding elements that influence the use of future bio-based medical devices and structures such as implants and stents. For example, the same biomaterial can be used to make cardiac stents and urinary catheters; however, the physical design will

Fig. 1.3 Countries represented by the biomaterials expert panel and a majority came from emerging nations (Greenwood et al. 2006)

be different to fit the particular need, and the material must be optimized to maintain bio-stability at the precise location.

The diverse collection of tissues and that exist inside of the body and the unique pathology of diseases have led to a wide range of contemporary biomaterials technologies. This growing field of science includes imaging agents, drug delivery platforms, biosensors, tissue engineered constructs, antimicrobials, and vaccines (Greenwood et al. 2006). The potential of biomaterials for tissue engineering and regenerative medicine to advance global health has been outlined by an international panel of 44 experts, including researchers and clinical workers in disciplines contributing to the production and application of regenerative medicine therapies (Greenwood et al. 2006). These scientists and clinicians, 77% of whom resided in developing countries, participated in a technology foresight study to identify the top ten most promising applications of regenerative medicine for improving health in developing countries (Fig. 1.3).

These applications range in scope from genetically engineered cells and DNA vaccines to bilayered skin constructs and growth-hormone-releasing dressings. These regenerative medicine applications aim to address and alleviate some of the world's leading causes of death and illness, including tuberculosis, pneumonia, HIV/AIDS, diarrheal diseases, and traumatic injuries. These applications of regenerative medical therapies reveal a potential opportunity for biomaterials to treat and alleviate both chronic and infectious diseases as well as injuries (Bhatia 2010). For many of these applications, in addition to effective treatment, biomaterial solutions can increase accessibility, while simultaneously reducing the cost (Greenwood et al. 2006). In the case of traumatic injuries, skin and tissue adhesives made from biomaterials can be used outside of the clinical setting to minimize blood loss and avoid the need for costly and often unavailable transfusions. The use and deployment of

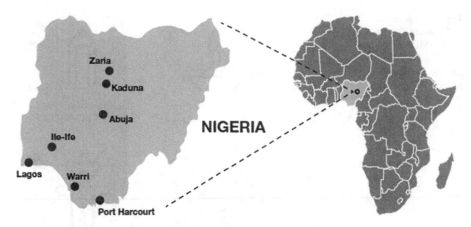

Fig. 1.4 Nigeria is located on the west coast of Africa along the Gulf of Guinea. Lagos is the most populous city in Nigeria and a major seaport (Dada et al. 2013)

biomaterials can facilitate innovative, accessible, and affordable treatment solutions for developing countries such as Nigeria.

Healthcare in Nigeria

The total expenditure on healthcare is represented as a percent of gross domestic production (GDP) in Nigeria at approximately 5.3%. To put this in perspective, the US total expenditure on healthcare is 17.9% of its GDP. Nigeria stands about average with other African countries, South Africa is slightly higher at 8.5%, and the Democratic Republic of the Congo is slightly lower at 2.5%. The life expectancy is approximately 53 years for males and 54 years for females (WHO 2013a). A major problem in developing countries such as Nigeria is not only the dissonance experienced in the quality of healthcare delivered, but also in their logistical support and access to medical infrastructure, supplies, and resources. The distance that patients in rural localities must travel in order to receive medical care in large urban cities, such as Abuja and Lagos, provides a huge barrier to adequate healthcare for many individuals (Stock 1983) (Fig. 1.4). While indeed access to healthcare facilities is a limiting factor to the delivery of adequate care, the services and outfitting of the healthcare facilities themselves is severely limited. For example, there are only 5 hospital beds per 10,000 Nigerians, 1 physician per 4000 Nigerians, and 1 healthcare worker per 1000 Nigerians (WHO 2013a). This limited access to properly outfitted physical infrastructure entities, such as healthcare facilities coupled with extremely limited human capital in the form of access to trained medical professionals, serves a detriment to many individuals that desperately seek medical care.

Fig. 1.5 The eight
Millennium Development
Goals (MDGs) serve as a
global action plan to meet
the needs of the world's
poorest (United Nations
2015)

In looking at what will be and has been done on an international level to mitigate the effects of the global burden of disease and human suffering, we look to that of the Millennium Development Goals (MDGs). There are eight Millennium Development Goals that represent the United Nations' blueprint effort to improve the lives of the world's most vulnerable and impoverished people, and several of them will be addressed throughout this case study (Fig. 1.5) (United Nations 2015). Each one of these aims are indeed daunting, but we will use Nigeria as an excellent example to demonstrate the potential for countries faced with pertinent health issues and lack of critical healthcare infrastructure to improve medical, environmental, and economic stability using renewable resources from their own territories. We investigate how some of these goals could be tackled with the introduction of bio-based materials from the Nigerian landscape. The materials of greatest interest include polymers extracted from corn and kenaf as well as phospholipids from soybeans.

Financial and Administrative Considerations in Nigeria

The amount of support available to address the health concerns of Nigeria is more pronounced than ever. In 2012, Nigerian President, Goodluck Jonathan, launched a bold initiative to save 1 million lives by 2015. Specifically, this effort aimed to expand access to essential primary care and healthcare services for Nigeria's youth and women. Of several components that will contribute to saving 1 million lives, there are four that directly relate to the focus of this case study (WHO 2012):

- Scaling up access to essential medicines
- Improving child nutrition
- Strengthening logistics and supply chain management
- Promoting innovation and use of technology

The backing, adoption, and implementation of these objectives by local government officials and members of the society are vital for promoting the sustainable future of healthcare access and delivery in Nigeria. As international momentum continues to increase toward achieving the Millennium Development Goals, we can see that the country of Nigeria is committed to leading the way for other countries to follow suit (WHO 2012).

In addition to support and increased awareness from leading officials, smaller groups have been formed in the past to help facilitate and guide healthcare development. TDR, the Special Program for Research and Training in Tropical Diseases, was formed three decades ago by three cosponsors: the United Nations Development Program, the World Health Organization, and the World Bank. Their vision, as depicted in their mission statement below, is similar in scope of what we seek to promote in our case study (WHO 2013d):

> "The power of research and innovation will improve the health and well-being of those burdened by infectious diseases of poverty." (TDR)

Furthermore, their mission demonstrates commitment to being a resource and sponsor for Nigerians as they transition into a new era that fosters the implementation of new technologies and innovations to improve the quality of life for all (WHO 2013d):

> "To foster an effective global research effort on infectious diseases of poverty and promote the translation of innovation to health impact in disease endemic countries." (TDR)

One important element to creating effective interventions utilizing innovations in any form is that of the ability to translate theory into effective interventions. The ability to take novel ideas and innovations and translate them in such a manner that creates feasible, efficacious interventions in the real world is vital. Often times this simple rhetoric of moving from theory to application is daunting. There are a multitude of confounding elements that make innovations that seem excellent on paper extremely difficult to carry out in real-world scenarios. In particular there are a multitude of barriers including fiscal capital, human capital, and political bureaucracy, to name a few. In applying novel innovations related to science, technology, and engineering applications in healthcare, this is often a dissonance in translating these innovations between developed and developing countries. For example, research, development, and production of advanced biomaterials are often carried out in highly developed countries that have access to a tremendous amount of human/intellectual and fiscal capital.

This dissonance becomes further relevant when medical technologies developed in advanced countries are only suited for operation in those countries with high infrastructure gradients. The limited adaptability complex of medical devices, and innovations in general, must be overcome. In countries such as Nigeria, the provision of healthcare services would likely significantly benefit from domestic manufactured bio-based materials that are created within their own specific systems, allowing for direct sourcing and economies of scale that can make these materials more fiscally accessible. The important step, and a challenge in implementation, is making the connection between innovation and financial resources to create the infrastructure necessary to see these innovations take to fruition. This is achieved through strong dialogue between policy-makers and scientific researchers, and these bridges can be crossed with support and mediations from groups like TDR.

In 2010, the TDR supported the launch of the ANDi operation, the African Network for Drugs and Diagnostics Innovation. The mission of ANDi is (ANDi 2010):

"To promote and sustain African-led health product innovation to address African public health needs through efficient use of local knowledge, assembly of research networks, and building of capacity to support economic development." (ANDi)

The director of TDR, Dr. Robert Ridley, explains that sustainable financing is an issue that taxes the minds of finance and health ministers (WHO 2010b). It is important to note that the ANDi operation is located within the United Nations Economic Commission for Africa (UNECA). This organization reports to the continent's ministers of finance and can count the African Development Bank as a valuable resource partner. Linking research on healthcare systems and biomedical innovation to economic development and poverty alleviation creates an integrative approach that can guarantee better access to sustainable funding efforts (WHO 2010b).

In addition to support from the Nigerian President, TDR, UNECA, and ANDi, biomedical research and development in Nigeria could expand if the country is chosen for the new home of the Initiative to Strengthen Health Research Capacity in Africa (ISHReCA) (WHO 2010b). Awarded more than $100,000 USD for operating costs by the Wellcome Trust's award, the ISHReCA is linked with ESSENCE

(Enhancing Support for Strengthening the Effectiveness of National Capacity Efforts) on Health Research to generate a focused dialogue between research and funders as well as provide a voice for African health institutions interested in capacity strengthening (WHO 2010b). These two groups potentially create a valuable avenue for financial and operational support for Nigeria to strengthen their impact on health. ESSENCE is a TDR initiative that works through the mounds of paperwork and conflicting priorities resulting from many new global health initiatives in an effort to harmonize funding programs and align them with the priorities of disease endemic countries (WHO 2010b).

Nigeria has demonstrated notable research success by the Nigerian National Institute for Pharmaceutical Research and Development (NIPRD) that developed the natural products-based formulation (Niprisan) from seeds, stems, fruits, and leaves of various African plants, for the treatment of sickle-cell anemia (WHO 2010b). Yet this solitary achievement in 1998 demonstrates that overall, current biomedical development falls short of addressing the country's most pressing health needs. Communication, collaboration, leadership, coordination, and financial sacrifice are the essential tools for research and development and have been lacking from Nigeria and the African continent as a whole (WHO 2010b). ANDi has made significant strides to improve these shortcomings resulting in a pledge from the African Union to increase its spending on scientific research and innovation to 1% of GDP, up from 0.3% of GDP in 2002 compared with the global average 1.7%. The European Union has also given ANDi 5 million euros in seed funding. These recent initiatives are summarized well by Ms. Aida Opoku-Mensah, Director of the Information and Communications Technologies (ICT) and Science and Technology Division at UNECA:

> "The continent of Africa is coming of age. We should learn how to turn our problems into solutions. We are too often looking for outsiders to come and help us. We have to do it ourselves." (WHO 2010b)

The support for healthcare remodeling and scientific innovation in Nigeria is widespread and active. The remainder of this case study will focus on identifying health priorities and address them with efficient and sustainable solutions.

According to the World Health Organization Neonatal and Child Health Profile, three of the leading causes of death among children under 5 years of age in Nigeria are pneumonia (20%), diarrheal diseases (15%), and traumatic injuries (4%) (Fig. 1.6) (WHO 2013b). Each one of these conditions could be improved with treatments extracted and engineered from sustainable, natural resources that can be found in or introduced to the Nigerian landscape. The Millennium Development Goal for under-five deaths in Nigeria is to reduce the mortality rate by 150% (Fig. 1.7). The following chapter uses specific examples to demonstrate how the evolving field of bio-based materials could be applied to address these critical pediatric conditions.

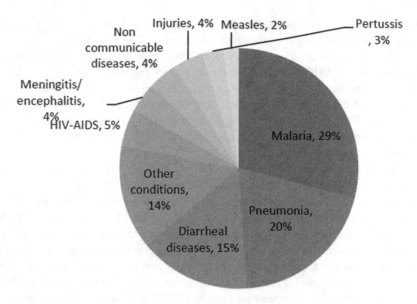

Fig. 1.6 Distribution of causes of under-five deaths in Nigeria (WHO 2013b)

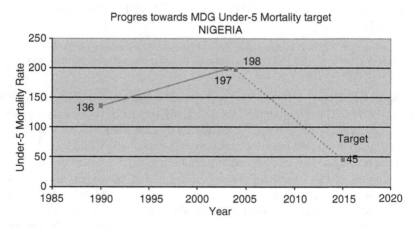

Fig. 1.7 Target reduction in under-five mortality rate for Nigeria based on the MDGs (WHO 2006)

Chapter 2
Biomaterials to Treat Pediatric Medical Conditions

Given the growing global burden of disease, novel treatment strategies are needed to address a multitude of health maladies and conditions. In acknowledging this, we use Nigeria as case study where feasible, functional interventions that utilize bio-based materials can be deployed infield at healthcare facilities to improve healthcare access, delivery, and most importantly, the quality of life for people. As previously mentioned, pneumonia, diarrheal disease, and traumatic injuries are among the top causes of morbidity and mortality for children under the age of 5 in countries such as Nigeria. In this chapter, each condition is discussed in terms of the clinical prognosis, current shortfalls in mechanisms for treatment, and the role-specific biomaterials could play in improving outcomes. The primary agricultural crops considered are soybeans, corn, and kenaf, for the introduction of biomaterials to manage these complex diseases, pathologies, and traumatic injuries.

Biomaterials: Effective Treatment of Pneumonia

Pneumonia and other lower respiratory infections are the third leading cause of death worldwide. Of the 4 million people who die every year from pneumonia, approximately 50% are children under the age of 5 (WHO 2008). Lower respiratory infections are the leading cause of death in developing countries and the leading cause of death for children under 5 worldwide (WHO 2008). In Nigeria, pneumonia is a leading cause of death in children under 5 accounting for 20% of total fatalities (Fig. 1.6) (WHO 2013b). A significant discrepancy exists between developed and developing countries with regard to childhood mortality from pneumonia: for every child that dies of pneumonia in an advanced country, 2000 children die of pneumonia in developing countries (WHO 2008). Antibiotics have been the primary treatment strategy against bacterial pneumonia; however, over time, multidrug resistant (MDR) pneumonias have become increasingly present, rendering many antibiotics

© The Author(s) 2018   15
A.A. Tracy et al., *Bio-Based Materials as Applicable, Accessible, and Affordable Healthcare Solutions*, SpringerBriefs in Materials, https://doi.org/10.1007/978-3-319-69326-2_2

ineffective (Stewart and Costerton 2001). The antibiotic resistance procured by pneumonia is the result of bacterial clusters forming drug resistant "biofilms" within the lower respiratory tract. These biofilms are a collection of microorganisms such as bacteria that form thin, glue-like films that make it difficult for antibiotics to penetrate the source of infection. Biomaterials may hold the key to penetrating these biofilms in order to destroy pathogenic microbes and reduce infection and mortality rates, but first we must explore the pathology of pneumonia to see how best biomaterials can be deployed (Meers et al. 2008).

In delving into the pathology of pneumonia, it is an inflammatory disease of the lung tissue that can result from a variety of infectious agents, including bacteria, viruses, fungi, and parasites. These can be introduced to the lower respiratory tract via inhalation of aerosolized material, by aspiration of upper airway flora, or by seeding of blood borne organisms from an infection in another part of the body (Dasaraju and Liu 1996). During the onset of pneumonia, the lung tissue becomes consolidated, alveoli become inflamed, and fluid exudate floods alveolar space. Lung defense mechanisms become suppressed or overwhelmed and mucinous material builds in the alveoli eventually leading to collapse of alveolar tissue. During an infectious event, the lungs lose their ability to absorb and transport oxygen into the bloodstream (Dasaraju and Liu 1996). In the case of Nigeria, pneumonia typically develops in an individual outside of a hospital setting and is classified as "community-acquired pneumonia." These cases arise when microorganisms residing in the nasophararynx are aerosolized into the lower airway (Dasaraju and Liu 1996). Common organisms associated with these infections are gram-positive bacteria, such as *Streptococcus pneumoniae* and *Haemophilus influenzae*, as well as respiratory viruses, including respiratory syncytial virus, adenovirus, influenza viruses, and parainfluenza viruses. Atypically, pneumonia can be caused by *Chlamydia pneumoniae, Mycoplasma pneumoniae, Legionella pneumoniae,* and zoonotic pathogens (Dasaraju and Liu 1996). It is important to focus attention on *S. pneumoniae* pathogenesis due to the fact that between 40% and 50% of all cases are caused by these bacteria (Speizer et al. 2006). Rapid recognition of the signs and symptoms, as well as effective treatment, is critical to saving individuals from acute pneumonia (Mulholland et al. 1992). Failure to treat severe pneumonia can result in death within hours as victims drown from fluid filled lungs (Dasaraju and Liu 1996).

The potential for biomaterials science to improve treatment strategies against pneumonia infections resides in developing drug delivery systems to attack and infiltrate bacterial biofilms (Bhatia 2010). These dense formations are the leading cause for antibiotic resistance and high mortality rates associated with pneumonia (Costerton et al. 1999). When pathogens such as *S. pneumoniae* infect lung tissue, they undergo a transformation in physiological state from planktonic or suspension form to a sessile or biofilm form (Fig. 2.1) (Monroe 2007). As the community of microorganisms grows and divides, a polymeric matrix coats the colonizing structure called the glycocalyx (Sutherland 2001). This hydrophilic and anionic polysaccharide secreted by the bacterial cells acts as a penetration barrier protecting the underlying infectious population (Sutherland 2001). Classical delivery of antibacterial agents fails to eradicate pulmonary biofilms as small-molecule antibiotics alone

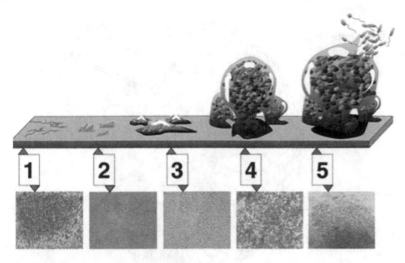

Fig. 2.1 Five stages of bacterial biofilm development: *stage 1*, initial attachment; *stage 2*, irreversible attachment; *stage 3*, maturation I; *stage 4*, maturation II; *stage 5*, dispersion. Each stage of development in the diagram is paired with a photomicrograph of a developing *P. aeruginosa* biofilm. All photomicrographs are shown to same scale (Monroe 2007)

are degraded by drug-inactivating enzymes and blocked by the glycocalyx from the source of infection (Smith 2005).

An alternative approach to pneumonia treatment is to introduce a submicroscopic carrier for the drug to hide and protect the molecule from reaction, degradation, or off target absorption (Bakker-Woudenberg et al. 1993). The most promising biomaterial for this purpose is liposomes (Fig. 2.2). These colloidal biomaterials are biologically, chemically, and physically stable and have been designed for controlled release of pharmaceuticals. What makes liposomes an attractive structure for targeting biofilms is their natural preference to localize at sites of inflammation and infection. They are biocompatible, biodegradable, and readily cleared by physiological metabolism. More importantly, they can carry, protect, and extend the circulation time of hydrophobic and hydrophilic drugs and are able to concentrate antimicrobial agents at biofilm interfaces (Jones et al. 1997).

The crucial ingredients of liposomal structures are amphiphilic phospholipids comprised of hydrophilic phosphate heads and hydrophobic fatty acid tails. When phospholipids are introduced to an aqueous solution, they spontaneously align based on their polar and nonpolar domains, generating a double-walled, hollow sphere that can encapsulate therapeutics. Phospholipids, such as phosphatidylcholine (PC) and phosphoethanolamine (PE), are major components of cell membranes but can be also be easily extracted from a variety of naturally occurring and available resources, such as the soybean (*Glycine max*) (Montanari et al. 1996) (Fig. 2.3).

Phospholipids are commercially available and can be obtained from the lecithin produced during the soybean oil refining process (List and Friedrich 1989). In 1989, List and Friedrich showed that supercritical carbon dioxide (SC-CO_2) was effective

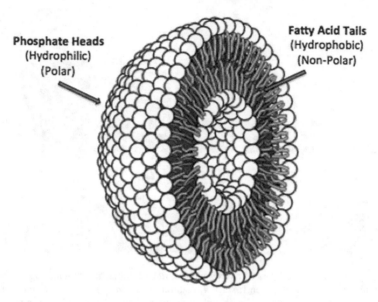

Fig. 2.2 Cross section of a liposome; key components are the phosphate heads and fatty acid tails (Bhatia 2010)

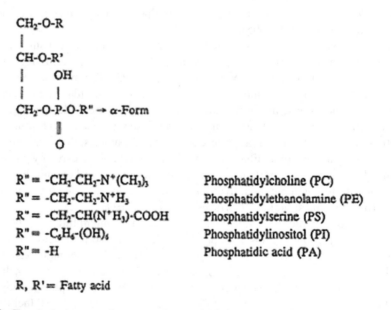

Fig. 2.3 Chemical structures of common phospholipids found in plants (Montanari et al. 1996)

in removing oils from seed matrices and is an appropriate process as the extracting agent is environmentally acceptable and nontoxic (List and Friedrich 1989). Montanari et al. further optimized this process in 1996 using SC-CO$_2$ and ethanol for the effective removal of oil and subsequent phospholipid fractions from soybean flakes for large-scale extraction (Montanari et al. 1996). Currently, Nigeria is the largest producer of soybeans in sub-Saharan Africa and exports approximately 20,000 tons annually (IITA 2009b). The by-products of this industry are essential sources for phospholipid production, and these ignored resources could have a tremendous impact on their approach to treating diseases such as pneumonia. Liposomal-encapsulated antibiotics could be especially useful in developing countries that are often plagued with resource deficiencies and limited healthcare infrastructure. The possibility of treating pneumonia with a single dose of antibiotics using phospholipid liposomes from soybean by-products would harbor significant benefits. Not only could it improve clinical outcomes, reduce the functional burden of disease, and reduce mortality rates in children, but also it creates an opportunity to increase patient efficacy rates for treatment. Since these are single-dose regimens, individuals could efficiently abide to the course of treatment and would not be hindered by the often long-term antibiotic treatment regimen that is often easily neglected as well as expensive financially.

The bioengineering process, such as transforming soy into a material suitable to aid in the treatment of pneumonia is quite obviously complex. The scientific details are difficult to scale up, and the financial implications are immense. However, we are here to show that limiting material dependency, particularly in the healthcare field, is vital to promote innovation in countries such as Nigeria in order to develop novel solutions to decrease the functional burden of disease that plagues so many.

Biopolymers and Traumatic Injury Treatment

As we are all aware, violence has plagued humanity since the dawn of time, and human transgressions are very much present in today's world. While indeed a tenet of the human condition is being innately altruistic to our fellow man, violence nonetheless knows no bounds. It is omnipresent in every society regardless of social or economic class, and the country of Nigeria is no different. In a recent report by the Institute for Economics and Peace, Nigeria was ranked as the seventh country most affected by terrorism (Institute for Economics and Peace 2012). The 2012 index was calculated based on four factors; the number of terrorist incidents, the number of deaths, the number of injuries, and the level of property damage. According to the records, in the year 2011, there were 168 terror attacks in Nigeria, accounting for 437 deaths and 614 injuries. The escalation of violence in the last decade is mainly due to a conflict between the Islamist militant group, Boko Haram, and the Nigerian Christians and also backed by weak central government of Nigeria. The Boko Haram uses violence and terror aimed at government offices, churches, and schools to destabilize the secular nation and is responsible for over 3000 deaths in northern and central Nigeria since 2009. In addition to religious tensions, Nigeria has witnessed violence arising from economic instability. After oil prices soared in early 2008, corruption in the government and rebel attacks on oil installation led to a dramatic fall in oil production, generating an economic collapse and subsequently increased violence (BBC News Africa 2013). It is this violence that has proliferated bodily harm for many people including an array of traumatic injuries that are often treated with the most basic of healthcare interventions.

In addition to injury caused by violent crimes and regional conflict, road traffic accidents are becoming increasing common in Africa. Sub-Saharan Africa has the world's smallest number of motor vehicles but the highest rate of road traffic deaths, and Nigeria is leading the pack with an estimated 53,339 fatalities in 2010 (Orekunrin 2013; WHO 2013c). A population-based survey in 2009 predicted that as many as 200,000 may have been killed as the result of road traffic accidents in Nigeria (Labinjo et al. 2009). Trauma has become an epidemic in sub-Saharan Africa, and the World Bank projects that the number of deaths will more than double by 2030 (World Bank 2014). If you add these numbers together with injuries due to rising communal conflict, it is clear that there is a significant public health emergency. This source of human suffering is a direct source of physical and emotional pain for a majority of the Nigerian population, and currently traumatic injuries account for 4% of deaths in children under the age of 5 (Fig. 1.6) (WHO 2013b).

The nature of these accidents and malicious attacks generate a wide variety of injuries to the victims, including burns, internal and external tissue trauma, and bone fractures (Rafindadi 2000). Treating such traumatic injuries is a challenging task for many reasons. Hospital and healthcare facilities are scarce and often lack the basic equipment and materials required to reduce blood loss and infection. Novel wound closure techniques are required to address a multitude of traumatic injuries in the field and without medical experts. For the victims that are fortunate enough to make it to the hospital, there must be readily available and sustainable materials for treating traumatic and surgical wounds efficiently. The optimal materials for wound closure and bone trauma must be biocompatible, nontoxic, supportive, and leak-free, minimize blood loss and infection, preserve tissue and organ function, and subsequently give the patient the chance of survival (Bhatia 2010). The key issue with treating traumatic injuries is time. In emergency situations the main concern is stopping blood loss. The materials used in these situations should be rapidly and reproducibly delivered, require minimal advance preparation time, practical in the field, and available at a moment's notice. Advancements in agricultural and biomaterials science have led to a number of potential materials that can aide in the delivery of emergency care, treating traumatic injuries, and improving the quality of life for individuals in countries such as Nigeria.

One of the most promising biopolymers that is receiving a significant amount of attention in the development of biomaterials for medical applications is polylactic acid (PLA). This naturally polymer can be produced by fermentation or chemical synthesis and extracted from agricultural resources, such as corn (*Zea mays*) and sugarcane (*Saccharum sp.*). It is being widely used in the biomedical area due to its excellent biocompatibility, biodegradability, and mechanical properties as well as broad scope of application in tissue engineering and drug delivery systems (Fig. 2.4) (Lasprilla et al. 2011; Xiao et al. 2012).

Théophile-Jules Pelouze first synthesized PLA in 1845 by the condensation of lactic acid. In 1932, Wallace Hume Carothers and colleagues advanced the process

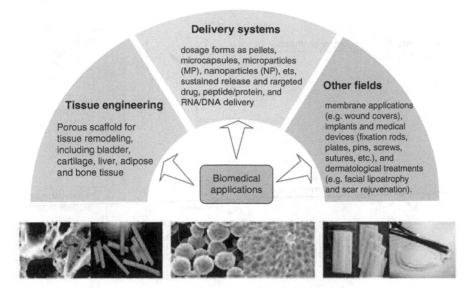

Delivery systems

dosage forms as pellets, microcapsules, microparticles (MP), nanoparticles (NP), ets, sustained release and rargeted drug, peptide/protein, and RNA/DNA delivery

Tissue engineering

Porous scaffold for tissue remodeling, including bladder, cartilage, liver, adipose and bone tissue

Other fields

membrane applications (e.g. wound covers), implants and medical devices (fixation rods, plates, pins, screws, sutures, etc.), and dermatological treatments (e.g. facial lipoatrophy and scar rejuvenation).

Biomedical applications

Fig. 2.4 Various biomedical applications of PLA (Xiao et al. 2012)

by developing a method to produce PLA via the polymerization of lactide, a synthesis technique that DuPont later patented in 1954. For decades the high price of PLA polymers and their low molecular weight limited its use to biomedical applications, such as biocompatible sutures and implants. A significant breakthrough occurred in the 1990s when Cargill Dow LLC successfully began to produce, at large-scale, high molecular weight PLA polymers under the name NatureWorks ™ (Auras et al. 2011). The life cycle of their PLA production begins with corn, a starch-rich, widely available, and renewable material (Fig. 2.5) (Vink et al. 2003).

The advantage of using a natural, agricultural entity, like corn, as a starting material over chemical synthesis techniques, is that much of the molecular synthesis has already been done by nature. The true challenge lies in separating the desired substance from all of the other components. During the process of photosynthesis, water and carbon dioxide use solar energy to make carbohydrates, primarily sucrose and starch, with a by-product of oxygen (Vink et al. 2003). This means that all of the carbon, hydrogen, and oxygen atoms that give rise to starch molecules originate from two simple components, water and carbon dioxide. Once the harvest is complete, the corn wet milling process can separate corn into its four major components: starch, germ, fiber, and protein. Cargill Dow converts the starch into dextrose, via enzymatic hydrolysis, and subsequently into lactic acid monomers through a series of fermentation and purification steps (Vink et al. 2003). Removing water from the lactic acid, monomer produces a low molecular weight prepolymer that can be depolymerized into lactide, a chiral cyclic intermediate dimer. The final step involves a solvent-free ring opening polymerization of lactide resulting in high molecular weight PLA. By controlling the purity and amount of each lactide stereoisomer, Cargill Dow has been able to make a variety of PLA polymers with unique

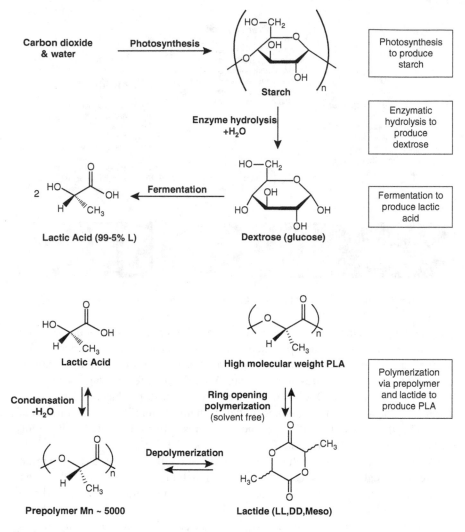

Fig. 2.5 PLA manufacturing overview established by Cargill Dow (Vink et al. 2003)

physical properties. Their commercially available compostable applications include, bedding, fiberfill for pillows, hygiene products, bottles, apparel, and wrap for consumer goods packaging to name a few (Vink et al. 2003). Cargill Dow has demonstrated the wide range of products that can be made from a crop such as corn.

In Africa, Nigeria is the largest producer of corn, accounting for 8 million tons annually (IITA 2009a). It is a staple food for the over 1.2 billion people in sub-Saharan Africa; however, the non-edible crop residue, such as leaves, husks, stems, and straw, can be used to make bio-based materials out of PLA to aid victims of traumatic injuries. It is a completely biocompatible material that is hydrolyzed by the body to form lactic acid, a normal product of muscle contraction, and subsequently

Fig. 2.6 3D-printed PLA Army-Navy surgical retractor (Rankin et al. 2014)

converted to water and carbon dioxide via the citric acid cycle (Athanasiou et al. 1998). For uses in the field, PLA could be formed into fibrous gauze pads or sleeves to cover an external wound or a burn, molded into stiff and protective splints for immobilization, or fabricated into a stretchy polymer sheet to wrap around the wounded area to provide pressure and reduce blood loss and inflammation. In the clinic, wounds can be treated using biodegradable polyglycolic sutures, which are completely self-dissolving and display biologically inert characteristics to limit unpredictable tissue reaction and reduce secondary inflammation.

A more recent development in the clinical applications of PLA has been in that of three-dimensional printing. PLA has the unique capacity to be utilized as a three-dimensional printing filament in a variety of printing apparatuses. Many pilot projects have been deployed in utilizing 3D printers outfitted with bio-based materials such as PLA to fabricate medical devices and supplies in field and on-site in medical facilities. PLA has recently been used to fabricate an array of medical tools and devices including vascular clamps, umbilical cord clamps, surgical retractors, and hemostats in the surgical field (Fig. 2.6) (Bhatia and Ramadurai 2017). PLA's unique materials properties including biocompatibility, high structural and torsional rigidity, and limited nanoparticle aerosol emission are ideal for surgical instrument fabrication. In addition, PLA is a low-cost and widely available filament, in which PLA-based surgical instruments can be fabricated for a fraction of the cost of conventional stainless steel instruments, while maintaining ideal instrument functionality and utility (Bhatia and Ramadurai 2017). Each one of these PLA applications in various medical supplies serves to improve patient outcomes and increase the interventional capacity for healthcare workers to delivery life-saving care for individuals afflicted with traumatic injuries. While indeed the applications of PLA in treating traumatic injuries are numerous, these applications must be thoroughly vetted and deployed in the field to test the efficacy against current options in the medical supply marketplace.

PLA alone can be fabricated to generate a variety of applications that can assist in the treatment of a multitude of injuries; however, more complex cases, such as those involving bone damage, may require different or supplemental biomaterials. Kenaf (*Hibiscus cannabinus*) is one such bio-based material that is gaining attention due to its excellent mechanical and chemical properties profile as well as its increasing cultivation worldwide (Avella et al. 2008). Kenaf's lignocellulosic fibers can be combined with PLA to form a biocomposite with increased strength and durability (Fig. 2.7). The major benefits of using kenaf fiber as an additive in PLA biomaterials are low density, low cost, nonabrasive nature, increased mechanical and thermal properties, and biodegradability (Bhatia and Ramadurai 2017; Sanadi et al. 1995; Ochi 2008).

For the past 6000 years, kenaf has been used as a cordage crop to produce twine, rope, and sackcloth, and its commercial use continues to diversify to include new applications such as paper products, building materials, absorbents, livestock feed, and biomedical scaffolds (Webber and Bledsoe 2002). Toyota Motor Corporation has even introduced high strength kenaf fiber blends into door panels to reduce weight and increase biodegradability (Ogbomo et al. 2014). Kenaf is a warm season annual fiber crop composed of stalks, leaves, flowers, and seeds and is closely related to cotton, okra, and bamboo. Major advantages of the crop include, short harvesting time, high yield, and its unique ability to be successfully grown in a wide range of soil types. Optimal growth occurs in well-drained and fertile soils; however, it can withstand flooding and drought conditions (Webber and Bledsoe 2002). The high protein content of the kenaf leaves make them ideal for animal feed, and the stalk can be broken down to produce low density fibers with high-specific strength (Webber and Bledsoe 2002). Bast and core fibers of the stalks can be blended with PLA to enhance properties important for biological purposes such as heat resistance, mechanical performance, durability, and moldability (Fig. 2.8) (Avella et al. 2008).

The addition of kenaf fibers to PLA biomaterials introduces a variety of biomedical materials for applications involving complex structures and functional strength. High fiber content PLA can be molded into sturdy scaffolds to support damaged bone or generate numerous prosthetics, and decreasing the amount of fiber could produce a softer scaffold to replace damaged cartilage. In addition to kenaf fiber reinforced PLA, soybean could also be used to repair damages and defects in the bone resulting from traumatic events. Soybeans are an attractive source for bone repair because they contain bioactive phytoestrogens that can stimulate differentiation of osteoblasts, bone-forming cells (Santin et al. 2007). The synthesis process involves thermosetting of defatted soybean flour, resulting in a biomaterial hydrogel that can be molded into films, membranes, porous scaffolds, and granules for surgical procedures (Santin et al. 2007). Soybean-based biomaterials have been combined with a number of composites to generate injectable foamed bone cements for minimally invasive procedures that increase the activity of bone-forming cells (Perut et al. 2011). The use of soybean-based injectables could be used to fill minor vertebral fractures and also as fixative for kenaf-PLA implants.

Fig. 2.7 Combination of kenaf fibers and PLA increases structural properties: (**a**) SEM of kenaf fiber bundle, (**b**) photograph of PLA resin, and (**c**) strength of material are dependent on fiber (%) (Ochi 2008)

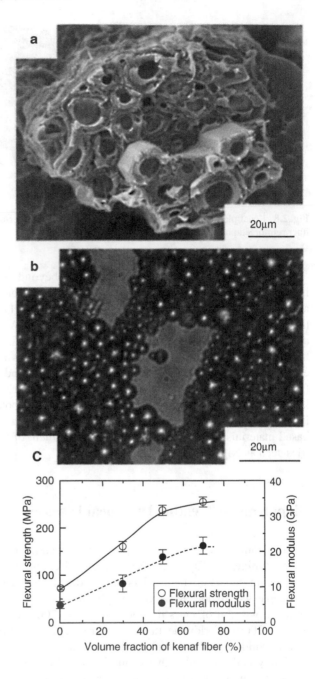

Fig. 2.8 Kenaf plant (left), kenaf stalk, kenaf fiber, and kenaf fiber-reinforced PLA (right) (Kenaf-fiber reinforced polylactic acid)

Natural fibrous PLA blends indeed extend beyond the use of kenaf fibers and include natural materials such as bamboo, flax, silkworm, and jute fibers (Bhatia and Ramadurai 2017; Li et al. 2013; Shih and Huang 2011). Figure 2.9 below shows the benefits of blending various natural fibers with PLA, in which these blends are also ideal for enhancing the mechanical properties of PLA. Specifically, these fiber blends can enhance PLA's overall materials properties including tensile strength, elongation break, bending and impact strength, and elasticity and ductility (Ramadurai and Bhatia 2017). As we can see, such materials ranging from corn and kenaf to soybean and jute materials possess the unique potential to be utilized either individually or in combination, illustrating the versatility and capability of bio-based materials to potentially enhance the treatment of traumatic injuries in countries such as Nigeria.

Biopolymers: Treating Diarrheal Diseases

While fatalities from traumatic injuries are spread far and wide and are perhaps the most prominently displayed cases, the global community is often unaware of the devastating impact of silent killers such as diarrheal disease, which plague millions of people worldwide. Rose George characterized diarrhea as "the world's most effective weapon of mass destruction" in his 2009 report *The Politics of Toilets* (George 2009). Worldwide, there are approximately 1.5 billion annual cases of diarrhea in children under 5, making diarrheal diseases the leading cause of pediatric morbidity and mortality. Over 2 million children die each year of diarrhea, one every 15 s, with a majority of cases arising from fecal contamination of food and water (Black et al. 2003). In Nigeria, diarrhea accounts for 15% of deaths in children under the age of 5 (WHO 2013b). The major problem with treating such a devastating disease is the inability to identify the underlying pathogen solely based

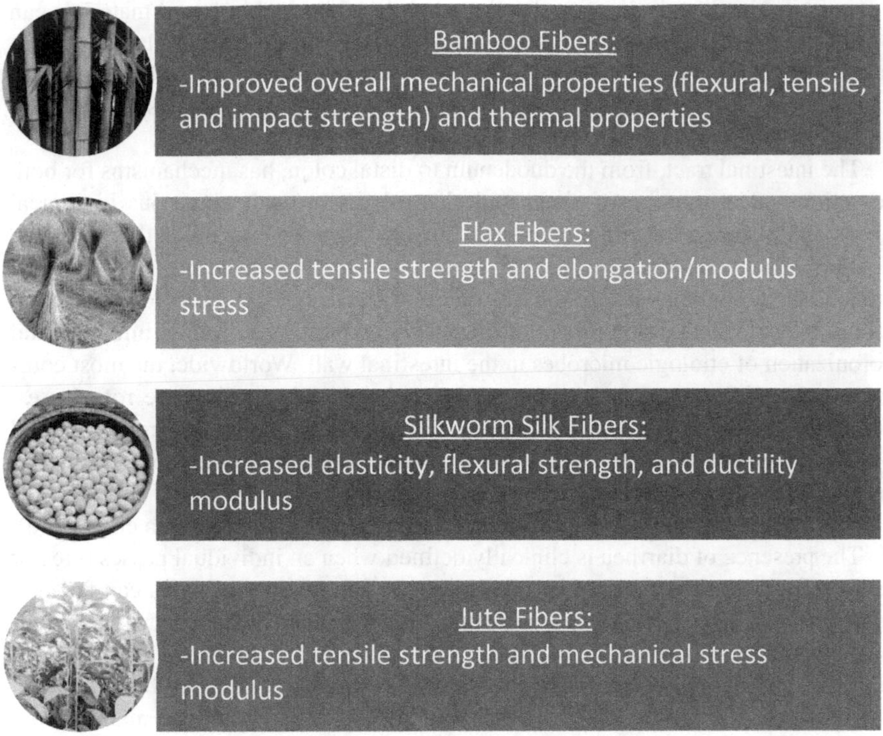

Bamboo Fibers:
-Improved overall mechanical properties (flexural, tensile, and impact strength) and thermal properties

Flax Fibers:
-Increased tensile strength and elongation/modulus stress

Silkworm Silk Fibers:
-Increased elasticity, flexural strength, and ductility modulus

Jute Fibers:
-Increased tensile strength and mechanical stress modulus

Fig. 2.9 Types of fiber biocomposite materials for PLA enhancement (Bhatia and Ramadurai 2017; Li et al. 2013; Shih and Huang 2011; Xiao et al. 2012)
Bamboo Picture: McMathis, J. (2014, August 22). Could a bamboo fiber composite replace steel reinforcements in concrete? Retrieved September 9, 2017, from http://ceramics.org/ceramic-tech-today/biomaterials/could-a-bamboo-fiber-composite-replace-steel-reinforcements-in-concrete
Flax Picture: Cuyler, S. (n.d.). Flax fiber-linen. Retrieved September 8, 2017, from https://www.emaze.com/@ALCQROOZ/Flax-Fiber-Linen
Silk Fiber Picture: Creative Commons Zero. (2016). Cocoon sliced silk brown silkworm white fiber – Max Pixel. Retrieved September 10, 2017, from http://maxpixel.freegreatpicture.com/Cocoon-Sliced-Silk-Brown-Silkworm-White-Fiber-722618
Jute Fiber Picture: Jute cultivation information detailed guide. (2014). Retrieved September 8, 2017 from http://www.agrifarming.in/jute-cultivation/

on clinical symptoms. There are a variety of microorganisms that can cause similar diarrheal symptoms, including viruses, bacteria, and parasites (Marshall 2002).

To address the dilemma of identifying the cause, biomaterials could be used to make low-cost diagnostic devices (Yager et al. 2006). Diagnosis of diarrhea often requires laboratory testing and healthcare infrastructure, which are generally not feasible in low-resource settings (Fatunde and Bhatia 2012). The use of sustainable biomaterials such as PLA can be employed in fabricating low-cost microfluidic devices that come preloaded with reagents to instantaneously and effectively diagnose diarrheal illness in patients. Deployment of the microfluidic devices in healthcare facilities in developing countries such as Nigeria can usher a paradigm shift in how diarrheal illness is diagnosed and treated. In addition, bio-based materials can be utilized in creating potable water infrastructure, a concept we will elaborate upon further in this chapter. In understanding how biomaterials and microfluidic devices can influence diarrheal illness prognoses in individuals, we must first explore the pathology of illness.

The intestinal tract, from the duodenum to distal colon, has mechanisms for both secretion and absorption of water and electrolytes to meet numerous biological needs and maintain intestinal homeostasis. More than 98% of fluids that enter the duodenum are reabsorbed in a healthy human intestine; however, during a diarrheal episode the normal net absorption of water and electrolytes is reversed to secretion (Keusch 2001). The development of diarrheal symptoms can follow infection and colonization of etiologic microbes in the intestinal wall. Worldwide, the most common and well-known enteric pathogens associated with diarrhea are rotaviruses, such as *Salmonella*, *Shigella*, and *Escherichia coli*. These pathogens are transmitted by a process known as fecal-oral transmission, where microorganisms from the stool of an infected individual spread to the mouth of others, with the degree of illness determined by the number and type of microbes ingested (Keusch et al. 2006).

The presence of diarrhea is clinically defined when an individual passes three or more sufficiently liquid stools in a 24-h period. The severity of cases can fall into three different categories: acute watery diarrhea, which may lead to varying degrees of dehydration; bloody diarrhea, which can occur due to intestinal damage as a result of inflammation; and persistent diarrhea, which is indicated by symptoms lasting 14 days or longer (Keusch et al. 2006). Poverty, environment, malnutrition, and young age are all factors that place and individual a higher risk for contracting diarrhea. Children in poverty-stricken areas have an increased risk as they are exposed to dirt floors, crowding, cohabitation with domestic animals that may carry infectious pathogens, lack of refrigeration for storage of food, and the lack of access to potable water (Keusch et al. 2006). Treatment of diarrhea involves oral rehydration, nutrient replenishment, and antibiotic medication to target the underlying pathogen. In developing countries these strategies are difficult to administer due to the lack of adequate, available, and affordable healthcare (Bhatia 2010). The most compelling issues that biomaterials can address are early and accurate diagnosis and the burden of recurrent and new episodes of diarrhea.

Before discussing possible routes for novel diagnostic platforms developed from biomaterials, the current approach to diagnosis must be mentioned in order to pinpoint areas for improvement. The identification of the causative pathogen is paramount for determining an appropriate treatment strategy to halt disease spread and prevent short-term and long-term complications. Currently, there is no single diagnostic test able to detect the thousands of organisms implicated in infectious diarrhea, and the only way to identify the underlying microbes is through a complicated series of laboratory tests. The lab workup may include multiple stool samples, microbiological culture, microscopic examination, pH, fat, and electrolyte measurement, enzyme immunoassays, latex agglutination, and toxin assays (Bhatia 2010). Many of these tests require a significant amount of monetary investment and several days to complete and may not even determine the infectious cause.

Fig. 2.10 Microfluidic lab-on-a-chip device utilized for diagnostic testing (Fraunhofer Life Sciences Engineering Group 2008)

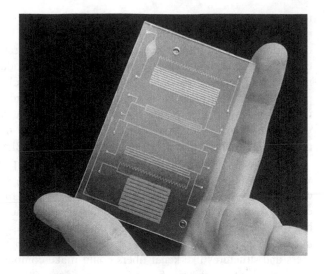

In countries such as Nigeria, there is clearly room for innovation and a novel approach to diagnosing diarrheal illness. When a country is handicapped by limited financial resources, inadequate equipment, and a lack of necessary laboratory personnel and facilities, it is imperative to create a low-cost diagnostic device that meets the needs of diarrhea patients and allows for immediate point-of-care diagnosis (Yager et al. 2006). Emerging biomaterials research is leading to unique and portable devices that are known as lab-on-a-chip devices (Fig. 2.10). These handheld chips function to miniaturize and automate complex laboratory procedures via microfluidics and could empower patients and healthcare workers with diagnostic information without requiring a central laboratory. Each chip made for the developing world must meet a variety of design constraints for optimal utility including, rapid and accurate screening, an easily interpretable interface that requires little training, and stability in variable environmental conditions (Yager et al. 2006). The theory and methodology behind microfluidic diagnostics are tiny volumes of sample and reagents, low energy consumption, faster analysis due to short diffusion distances, and precise manipulation of complex fluids without the need for an expert operator (Toner and Irimia 2005). These devices can be just a few square inches in size yet have the built in capacity to detect multiple pathogens via massive parallel reactions and measurements.

Microfluidic devices can serve a variety of purposes in many industries and are often fabricated with expensive components such as glass, quartz, and silicon. In order to minimize the cost of a novel diagnostic platform, inexpensive materials such as PLA biopolymers could be used as an alternative and microfabricated via injection molding for production of large numbers (Surace et al. 2012). A collaborative effort between the Program for Appropriate Technology in Health (PATH), Micronics, Inc., and labs at the University of Washington and the George Washington University has generated a diagnostic enteric card (DEC) for the simultaneous detection of six different enteric pathogens (Weigl et al. 2012). The technology takes advantage of distinct receptors on the bacterial surface for the immunocapture of each pathogen with a projected assay cost of $1–5 USD. This consortium has opened the door for biomaterials scientists by demonstrating successful identification of enteric pathogens using microfluidics. The production process, assembly method, and detailed reactions required to generate a clinical diagnosis given the pathogen complexity of diarrhea is a significant research effort to address. Novel rapid diagnostic devices using biomaterials, like PLA, will empower healthcare workers in Nigeria to provide timely and appropriate treatment for patients and reduce the amount of lives lost from diarrheal illness (Bhatia 2010).

The practice by farmers to distribute noninfectious products to the community via diagnosis through lab-on-a-chip devices is one way to show how biomaterials can indirectly have an impact on human health. Ideally, the goal should be to eradicate the source of disease before individuals are made victims. Biomaterials science has the potential to decrease the global burden of disease, in particular the disease burden of diarrhea, by introducing preventative technologies. For example, biopolymers and structures like kenaf fibers could make complex networks coupled with nanoparticles for water purification systems (Juma 2010). Nearly 300 million people in Africa lack access to clean water. "Smart plastic" membranes derived from agricultural resources could increase sanitation while decreasing the number of infections. This process is also half the cost of traditional purification methods like reverse osmosis, requires 30% less energy, and does not involve any toxic elements (Juma 2010). Once purified, water could be stored for extended periods of time in large containers or bottles manufactured out of PLA to provide easy and available clean water sources for communities and individuals. Being generated by biomaterials, these bottles would be compostable without negatively affecting the surrounding environment and recycled often to maintain clean storage. Similarly, engineered toilets, like bedpans, could be used to properly dispose of stool in areas away from sources of water and resources.

Not only can certain biopolymers be degraded to their original form for subsequent production of diverse materials, but they could also be returned to the land for increased agricultural productivity. A paper recently published by the Nigerian Society for Experimental Biology reported that introducing various cross-linked starch polymers not only supports crop growth but also provides cleaner water runoff by eliminating 80–99% of the sediment (Ekebafe et al. 2011). These biopolymers can also be used as vehicles for controlled release of biocides and herbicides into the soil. This effect increases efficiency while reducing cost and toxicity by lowering the amount of chemicals needed (Ekebafe et al. 2011). Providing Nigerians with clean water and limiting their exposure to harmful pathogens from a myriad of cross-contamination scenarios are necessary steps to reduce the burden of diarrheal diseases.

The challenges that lie ahead for Nigeria to adequately combat diarrhea are indeed numerous and rely heavily on concentrated efforts and the proper allocation of resources. Diarrheal illness is a multifaceted illness that not only impacts the immediate health of individuals but also impacts the social and economic fabric of entire countries. The treatment of waterborne diseases like diarrhea cost

Table 2.1 Summary of potential outcomes with the use of biomaterials for three pediatric medical conditions

Condition(s)	Under-five deaths	Crop(s)	Biomaterial(s)	Product(s)	Outcome(s)
Pneumonia	20%	Soybean	Liposomes	Drug delivery vectors	Localized and controlled therapeutic release
Traumatic injuries	4%	Corn, soybean, Kenaf	PLA, hydrogels	Wound closure products, splints, prosthetics, injectable	Reduced blood loss, cartilage and bone repair, proper healing
Diarrheal disease	15%	Corn	PLA	Diagnostics, water and soil filters, storage containers	Increased sanitation, improved water sources, reduction of harmful runoff

sub-Saharan African governments at least 12% of their total health budgets (UNDP 2006). Early malnutrition from diarrhea negatively affects growth and development of children and has long-term effects on the overall health of populations. The need for novel diagnostic devices is intensified as antibiotic-resistant pathogens continue to emerge (Bhatia 2010). Agricultural biomaterials have the potential to overcome a portion of the destruction caused by diarrhea through preventative measures as well as improved diagnostic technologies. The bio-based materials and microfluidic devices discussed in this chapter can serve as powerful resources and elements of change that could improve a variety of health outcomes (Table 2.1).

Chapter 3
Impact of Biomaterials on Health and Economic Development

Improving the medical outcomes for pneumonia, traumatic injuries, and diarrheal diseases extends beyond the overall health of affected individuals. The *health* of a country is innately tied to the corresponding *wealth* of that country, in which healthy individuals foster increased productivity and economic gains on both local-level and macro-level scales. In acknowledging this, there are additional steps that Nigeria can consider in the area of biomaterials science that may lead to greater health and economic sustainability. The primary concern is the health of future generations in combating conditions and morbidities that afflict the youth population and human capital development and minimizing the risks associated with increased economies of scale related to the future development of bio-economies. Biomaterials have the potential to foster advancements on multiple fronts but require an integrative approach to making sure that future initiatives are both sustainable in nature as well as help those that are in need of these interventions the most.

The Burden of Pediatric Conditions: A Biomaterials Approach

This case study has demonstrated the significant role that biomaterials could play in improving health outcomes for three of Nigeria's most serious pediatric maladies. In 2012, there were over 800,000 under-five deaths and together, in which pneumonia, diarrheal disease, and traumatic injuries accounted for 39% of these losses (WHO 2013b). This extraordinary number does not take into account the millions of children who are currently suffering and living with complications associated with these conditions. This section will review the clinical implications of each disease and injury, highlight detrimental secondary effects, and discuss the overall economic impact of childhood morbidity and mortality.

In the case of pneumonia, affected children can have severe and prolonged infections that may lead to chronic lung damage. These infections may form a lung

© The Author(s) 2018

A.A. Tracy et al., *Bio-Based Materials as Applicable, Accessible, and Affordable Healthcare Solutions*, SpringerBriefs in Materials, https://doi.org/10.1007/978-3-319-69326-2_3

abscess, and as more lung tissues are destroyed, pneumonia can cause respiratory failure and acute respiratory distress syndrome (Mulholland et al. 1992). Children who suffer from pneumonia often carry a poor prognosis, and those who survive experience episodes of recurrent pneumonia and restrictive lung disease through mid-adulthood and beyond (Edmond et al. 2012). Not only can pneumonia have long-term effects on the lung, but it can also develop into systemic septicemia resulting in liver, kidney, and heart damage (Bhatia 2010). These complications put children at high risk for secondary effects, such as decreases in weight and inability to participate in physical activities. The use of liposomal biomaterials to increase the antimicrobial effect of treatments could significantly decrease the burden of disease both in the short and long terms.

Second to pneumonia as the leading killer in children worldwide, diarrheal diseases cause more morbidity than any other ailment (WHO 2008). This illness is increasingly recognized as a widespread health problem with complications that are both acute and long term. Multiple repeated enteric infections are devastating for children whose diets are marginal and can be especially problematic in achieving normal nutrition (Guerrant et al. 2008). Diarrhea can potentially lead to lifelong disability, and subsequent malnutrition impairs weight and height gains, as well as deficiencies in both physical and cognitive developments (Guerrant et al. 2008). In addition, malnourished children are predisposed to both increased incidence and duration of diarrheal episodes, leading to a vicious and perpetual cycle (Guerrant et al. 2008). The Water and Sanitation Program estimates that nearly 90% of deaths from diarrhea are directly attributed to poor water, sanitation, and hygiene. Collectively, the burden of diarrheal diseases and associated malnutrition causes Nigeria to lose up to $3 billion USD per year or approximately 1.3% of national GDP (Sittoni and Maina 2012). It may take an extensive period of time and a substantial effort to make the necessary changes to overcome the sanitation issues in countries such as Nigeria. However, biomaterials could begin to deliver improved outcomes at both the human health and environmental levels. Novel filtration devices derived from natural resources would improve access to clean water, and recyclable latrine units would give the urban populations a way to appropriately dispose of waste (Juma 2010). New biomaterials for rapid, accurate, low-cost, point-of-care diagnosis of the causative pathogens could potentially have an immediate impact where caregivers can provide appropriate treatment (Yager et al. 2006).

Contrary to the previous two health conditions, there is no diagnostic tool or medical treatment that can assist in reversing the wounds caused by traumatic injuries. Permanent scars are often left behind, and amputated limbs lead to permanent disability; this is still a reality despite advancements in suturing and wound closing techniques, in which surgeons still continue to struggle with difficult tissue repair (Bhatia 2010). In a country where there is only one hospital bed per 2000 people, the focus for improved outcomes related to traumatic injuries in Nigeria must begin at the community level (WHO 2013a). Access to emergency medical supplies such as basic trauma surgical instrument kits for individuals who suffer injuries from traumatic events is pertinent to enhancing the interventional capacities of primary care facilities. For children who experience severe injuries resulting in tissue loss, the road to recovery can be long and complicated. Such injuries leading to long-term disability and reduced quality of life are often overlooked in part because of the well-recognized burden of disease (Hofman et al. 2005). Customized splints and casts fabricated with on-site three-dimensional printing technology utilizing biomaterial filaments such as PLA could provide a low-cost option to providing care. The use of these sustainable, domestically fabricated biomaterial devices can assist in proper healing and could be the difference between normal growth and permanent disability.

In further exploring the applications of biomaterials in alleviating injuries, we can also look beyond those that are traumatic in nature. Developing countries have heavily agriculture-based economies, in which many people work in hard labor jobs related to farming, especially youths. The occupation that presents the greatest relative risk for developing osteoarthritis is farming (Woolf and Pfleger 2003). Osteoarthritis, the leading cause of disability, is characterized by loss of cartilage between joints that is clinically associated with joint pain, tenderness, limited movement, and varying degrees of local inflammation. Akinpelu and colleagues estimate that around 20% of adults aged 40 or older from rural areas of Nigeria have symptomatic osteoarthritis (Akinpelu et al. 2009). Silk is one biomaterial that can be used to lift the injury burden experienced by a large population working in physically demanding conditions, such as farmers. The toughness of silk fibers is greater than the best synthetic materials, and the biodegradability of silk fibroin protein makes it an excellent polymer for biomedical applications (Rockwood et al. 2011). As a biocompatible resource, silk fibroin can be formed into injectable gels to induce cartilage growth and repair for patients affected by osteoarthritis (Rockwood et al. 2011). Reducing the burden of injury for chronic pain with a biomaterial like silk could improve agricultural production and the well-being of the farmers that Nigeria greatly depends on.

Previous studies have demonstrated the economic burden of increased medical costs on developing countries (WHO 2013a; Bhatia 2010; Sittoni and Maina 2012). Catastrophic healthcare expenditures are an often-overlooked facet related to healthcare access and delivery in many countries. This occurs when relative medical costs are greater than or equal to 40% of a household's non-subsistence income. This can place the families of individuals who have been injured on path to financial disaster and impact intergenerational equity. This occurs in addition to the loss of productivity or development of an individual during the course of their ailment. If the individual that is the primary income source for a family is sick, this creates an even greater fiscal burden for those families to survive. As we previously mentioned, the health of a population is tied to its wealth and economic output/productivity. Sick people cannot work and thus cannot contribute to the outputs of an economic system. An interesting phenomenon is that even if perhaps stable healthcare infrastructure was in place and adequate medical supplies were provided, many individuals may elect to still not utilize essential surgical services to treat a serious ailment. Most individuals are not covered by health insurance and pay for healthcare services out of pocket and cannot afford either the direct costs of medical care including consultations and medications or the indirect costs including medical

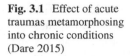 **Fig. 3.1** Effect of acute traumas metamorphosing into chronic conditions (Dare 2015)

transport. This serves as a focal barrier to effective treatment and access to healthcare for many and effectively contributes to the cyclical nature of poverty. This can eventually cause an acute trauma that could be easily remedied, such as a long bone fracture in a youth, to turn into a chronic condition that could create a permanent disability for the youth and impact their livelihoods in the future. Low-income households that are plagued by illness can progressively dive further into poverty due to the adverse effects of illness on their respective income and overall welfare as depicted in Fig. 3.1 below.

Oftentimes the indirect costs associated with death, illness, and injury are difficult to access and calculate. In particular, the indirect costs associated with child morbidity and mortality represent a loss in the future intellectual/labor output of these children. The factors associated with these costs include shifts in population dynamics, lost time trying to access primary care, decreased time in school due to injury or illness, and absent time from work caring for another, just to name a few. The Water and Sanitation Program estimates that, in Nigeria, these indirect costs can add up to $300 million USD (Sittoni and Maina 2012). The World Health Organization created a guide to identify economic consequences associated with disease and injury that was based on several case studies around the world in this specific area (Evans et al. 2009). One obvious trend that was observed in these studies was that the improvements in overall health and sanitary conditions resulted in significant declines in the rates of child mortality. Furthermore, as parents adapt to decreases in child mortality, the fertility rates also tend to decline (Evans et al. 2009). These demographic transitions accounted for approximately 1.4–1.9% of annual GDP per capita growth in East Asia during 1965–1990 (Bloom and Williamson 1998). Similarly, the United Nations Commission on Population and Development (UNCPD) estimates that fertility decline in Nigeria could have reduced poverty by 14% between 2000 and 2015 (Leke et al. 2014). The broader concept of economic welfare is often missing from these calculations, such as the value of leisure, knowledge, and health itself (Evans et al. 2009). These components are key to a prosperous and sustainable population. Investing in the development of biomaterials integrated into novel medical devices and technologies can create an impetus in improving health outcomes for devastating pediatric conditions in Nigeria and fuel future economic growth.

Human Capital Development: Closing the Educational Gap

During the course of this book, we have touched upon the fundamental concept of human capital and its relationship with public health as well as healthcare and medical device innovations. This is a concept that is often overlooked, particularly when dealing the global burden of disease. Oftentimes we focus on the immediate effects of disease and illness, this being reflected in morbidities and fatalities, but what about the intellectual capital development and future of individuals and their greater nations? Intellectual equity is a nonphysical asset that has the power to change the future of developing countries. Education is often touted as the key breaking the poverty trap, and when it comes to the implementation and interventional capacities of biomaterials to be implemented in medical device innovations, it is absolutely vital. The impetuses of innovation are rooted in our ability to utilize education as a foundation for creative discoveries and innovations. The youth of our nations hold the key to defining the innovations of tomorrow, and this is no different in Nigeria.

In order to build upon the population and health gains created by investing in the welfare of Nigeria's youth, the proper educational framework should be established for continued improvement. Well-educated and highly skilled workers are the building blocks of a productive economy, and Nigeria has a great deal to do to improve education and training (Evans et al. 2009). More than 10.5 million children between the ages of 6 and 17 are not in school, and even after 6 years in school, only 1 in 5 children aged 15–29 can read and write (Leke et al. 2014). In addition to low enrollment and completion rates, a distressing fact about many African school systems is the lack of focus on maximizing community-based resources and opportunities. Instead, they tend to focus on skills less applicable to village life that aim for the rare occasion of moving to rural areas (Juma 2010). The result of poor-performing Nigerian schools is an adult population that lacks the necessary skills for highly productive economy. In the country, there are more than 35 million adults who cannot read and write (Leke et al. 2014).

African leaders have a unique opportunity to use the agricultural system as a driver for their economies and the future of sustainable innovations. Most developing countries have economies that are largely agricultural-based, relying on hard labor to harvest and plant crops. In connecting this to human capital, individuals learn agriculture as a subject in formal schooling, from the earliest childhood experience to universities, in which populations could gain access to improved farming techniques, increased production methods, and deep technical knowledge to spur innovations in the agricultural segment (Juma 2010). But what does this have to do with biomaterials? As we previously discussed in Chap. 2, many natural biomaterials are derived from agricultural crops and products. Improving cultivation and synthesis of these biomaterials from these agricultural stocks can prime Nigeria to effectively synthesize and manufacture an array of biomaterials to be utilized in medical devices as well as a host of other applications. But agricultural systems must be linked to engineering systems, oftentimes, which require the most educated individuals and access to human capital.

Not only should Nigeria treat agriculture as a skill to be learned, valued, researched, and improved upon, they should use collaborative methods to connect its systems with the growing engineering sector (Dada et al. 2013). Sufficient engineering capacity is essential for economic, social, and technological development, the creation of improved infrastructure entities including healthcare, as well as cultivation of an attractive environment for foreign investments (Matthews et al. 2012). The historical view that growth and development can be facilitated by imported expertise has diminished as nations understand that local innovation and capability should be invested in and nourished so that solutions can be developed in-country (Matthews et al. 2012). Transforming the agricultural sector by incorporating engineering expertise could allow Nigeria to take advantage of their existing resources, ecological systems, and industrial capacities (Juma 2010).

A large amount of capital is needed to create these education, engineering, and improved agricultural systems. Most importantly investments in these various areas will prevent the phenomenon of "brain drain" that plagues many developing nations. Brain drain refers to the emigration of highly trained individuals from a country that often has underdeveloped or limited resources to more advanced countries. These advanced countries often harbor better paying jobs and positions that they individuals can be employed in, with far better outcomes than those if they stayed in their native countries. These highly trained individuals include doctors, scientists, lawyers, and engineers that are vital contributors to any economy. Developing countries often have diminished access to human capital and often have very few of these highly trained individuals. Incentivizing these individuals to stay in their native countries and contribute to such vital systems as healthcare infrastructure can ultimately improve not only the intellectual capacity for innovation but also the health and wealth of these countries in the future.

Connecting agricultural, engineering, and healthcare systems could potentially serve as an impetus to further research and design novel medical devices and applications for Nigeria's immediate and future healthcare needs. Currently Nigeria, and Africa as a whole, is behind in regional collaborative efforts to find solutions and improve healthcare access and delivery. A majority of research collaborations and innovations are carried out by the United States and Europe, and this represents a major challenge. This prevents Africans from having access to the intellectual resources and products of these collaborations, ultimately hindering their respective medical research agendas and future of fostering medical innovations in Africa (Vré et al. 2010). For example, malaria is the leading killer in children under 5 in Nigeria, and although the University of Ibadan has published between 15 and 30 research articles, they only have a single collaboration with another African Institute, in

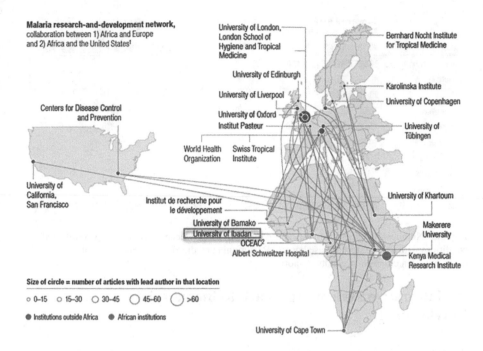

Fig. 3.2 A collaboration bias for the malaria research and development network. The University of Ibadan, Nigeria, falls behind several European countries in number of publications (Vré et al. 2010)

which their collaborations with European countries have produced far more publications (Fig. 3.2) (Vré et al. 2010). The same effect is seen for lower respiratory infections, like pneumonia, and diarrheal diseases, where the basic research output is clearly a misrepresentative of the burden these diseases cause African nations (Fig. 3.3) (Vré et al. 2010).

Ultimately, the solutions to these issues lie within Nigeria itself. Nigerians pride themselves on fostering a strong sense of community, in which Nigeria should continue to invest in human capital development and training the science, technology, engineering, and math (STEM) fields to produce future generations with the capacity to develop novel innovations not only related to medicine but a host of other fields. In addressing their healthcare needs, efforts must be made by Nigeria on a social, economic, and political level to enhance healthcare systems and access to care to individuals that need it most. In ushering a future of biomaterials utilized in Nigeria, a collective effort must be taken to develop a highly skilled workforce that can contribute to developing novel engineering sectors focused on creating novel bio-based medical devices and technologies. The cultivation of health professionals who are going to push the limits of innovation and take ownership of their respective research agendas is also vital.

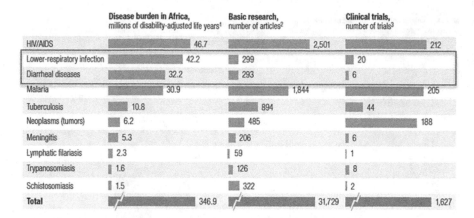

	Disease burden in Africa, millions of disability-adjusted life years[1]	Basic research, number of articles[2]	Clinical trials, number of trials[3]
HIV/AIDS	46.7	2,501	212
Lower-respiratory infection	42.2	299	20
Diarrheal diseases	32.2	293	6
Malaria	30.9	1,844	205
Tuberculosis	10.8	894	44
Neoplasms (tumors)	6.2	485	188
Meningitis	5.3	206	6
Lymphatic filariasis	2.3	59	1
Trypanosomiasis	1.6	126	8
Schistosomiasis	1.5	322	2
Total	346.9	31,729	1,627

Fig. 3.3 Number of research articles and clinical trials produced in Africa for several medical conditions with the greatest disease burden. Pneumonia and diarrheal diseases, two of the top killers in Nigerian children, are in the minority

An Integrative Systems Approach to Biomaterials Development

As one can see, developing and securing human capital will be an enormous, yet feasible, feat for Nigeria. Once major connections are made within the engineering and healthcare sectors, the true potential for growth and innovation will take to fruition and not go unnoticed. In creating a "bio-economy," demand for agricultural resources necessary to sustain this type of economy could cause a further significant boom in agricultural production that could pose certain risks. Given that many farmers utilize synthetic fertilizer, local environments could see excessive levels of nitrogen pollution, if indeed demand for agricultural products is increased in creating this bio-economy (Ogburn 2010). Reactive nitrogen, the chemical that helps improve crop yields, can escape the soil entering the atmosphere as a potent greenhouse gas and degrade the habitat of surrounding sources of water (Ogburn 2010). In order to reduce the amount of fertilizer required to increase growth, farmers should take advantage of recycling and reusing biomaterial polymer stock in order to control the release of biocides (Ekebafe et al. 2011). While there are indeed trade-offs associated with implementing any type of new economic structure, potential gains that are to be experienced from establishing a bio-economy in Nigeria are quite profound. Key elements that we have been outlining in this chapter are the three specific areas where biomaterials can play a major role in lifting the economic, social, and physical burden of disease. While we can identify the areas in which biomaterials seek promise in remedying the various facets of burden of disease, a unique approach must be fostered to feasibly ensure the propagation of biomaterials development in Nigeria. We term this as an "integrative systems approach" whereby we acknowledge that the future of biomaterials development for healthcare entities in Nigeria will rely on the complementary integration of three core systems, this

Fig. 3.4 Framework and feedback loops that are vital in addressing the healthcare needs of Nigeria

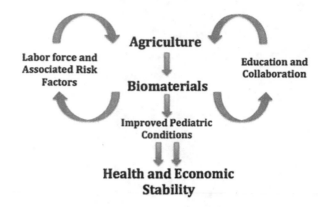

being education, agriculture, and domestic engineering systems. Together, education, research, and functional collaboration between domestic and international agencies can serve as a catalyst in securing the future health and wealth of Nigeria (Fig. 3.4).

Future healthcare systems in bio-economies will rely on a culmination of these elements, as they will rely on the innovations fostered by the use of advanced materials such as biomaterials derived from natural resources and agriculture, the engineers and scientists to create new medical devices, and lastly the human capital in the form of doctors, nurses, and other medical practitioners. The integration of these elements will surely strengthen healthcare access, delivery, and innovation in Nigeria and create a paradigm shift in how we see healthcare in the future. In tackling the multitude of pediatric conditions associated with the global burden of disease, we can secure a future for Nigeria's youth for generations to come as well as potentially have these youths further contribute to redefining healthcare systems and institutions. It is indeed the youth that hold the key to defining the trajectories that all countries will follow.

Chapter 4
Discussion and Conclusions

Biomaterials derived from natural resources have demonstrated significant promise for medical use in the modern era and have steadily been gaining ground in developing countries such as Nigeria. Development of "successful technologies" in the form of natural therapeutics and that utilize bio-based materials could help to potentially lower the barrier to adequate healthcare access and delivery in a multitude of ways. Locally designed and manufactured medical devices and supplies can be fabricated with biomaterials derived from natural, native resources and can be customized to further cater the healthcare needs of the people of Nigeria. Although the initial capital expenditure and investment can be quite high in order to build the infrastructure to support the generation and distribution of biomaterials, the decreased dependence on imported and donated medical supplies, devices, and products could lead to more cost-efficient and sustainable solutions in the future. The true test lies in making the key connections between agriculture, engineering, human capital development, and medicine which is an integral step in improving economic and healthcare outcomes in the future of Nigeria.

Our goal for this book was to highlight potential disruptive innovations to the current healthcare access and delivery paradigm that Nigeria is currently experiencing, as well as address some of Nigeria's greatest healthcare needs. The country's leaders and citizens undoubtedly recognize the problems related to their healthcare systems, which is hindering their ability to achieve economic and social prosperity. The support for healthcare remodeling and scientific innovation within the country is active, and research around biomaterial use and deployment in various medical settings should become a major effort for further investigation (WHO 2010b). As previously mentioned, a portion of the Millennium Development Goals as outlined below highlights a few specific areas where biomaterials science can have significant impacts (United Nations 2015).

© The Author(s) 2018
A.A. Tracy et al., *Bio-Based Materials as Applicable, Accessible, and Affordable Healthcare Solutions*, SpringerBriefs in Materials, https://doi.org/10.1007/978-3-319-69326-2_4

- Goal 1: Eradicate extreme poverty and hunger

Nigeria's landscape spans 225 million acres, and just over 200 million acres are believed to be arable. Interestingly only 42% of this cultivable land is actively farmed (Olomola 2007). These numbers strongly support the expansion of crops to be used for purposes other than food resources and settle any ethical issue that may arise regarding a depletion of natural resources for consumption and shelter (Helen 1991). As much as 70% of the Nigerian population works in the agriculture sector and would be vital in creating a new market niche such as a bio-economy in the future. In establishing the foundations for a bio-economy, agricultural products would be vital in fostering the raw material for bio-based materials and devices. These materials would be utilized by the engineering and healthcare sectors and could create a stable supply-demand relationship that promotes the creation of more jobs, open avenues for foreign investments and economic development, and generate new opportunities for export (African Progress Panel 2014; Olomola 2007). By expanding agricultural infrastructure and development to promote economies of scale to develop a bio-economy progressively, the overall health and wealth of surrounding local communities could be improved. Increased investments in agricultural infrastructure and processes to improve material delivery for a biomaterials market niche can promote higher yields, improve incomes/job creation, and ultimately reduce the burden of hunger. Improving education and training around best agricultural practices would allow smaller communities to locally establish useable, profitable, and sustainable sources of goods.

- Goal 4: Reduce child mortality and Goal 6: Combat disease

The child mortality rate in Nigeria, 124 per 1000 live births, is nearly triple the global average, 48, and much higher than the regional average, 95 (WHO 2014). In 2012, the WHO estimated that there were 826,604 under-five deaths (WHO 2013b). At this time, Nigeria is far above their target mark of reducing the mortality rate to a level closer to the global average (Fig. 1.7) (WHO 2006). There are multiple confounding factors that have contributed to this mortality burden, such as lack of access, insufficient treatment options, and poor quality care (WHO 2006, 2013b). The second chapter of this case study has shown that for three specific pediatric conditions, which account for a large portion of the mortality rate, various biomaterials could play a role in improving pediatric outcomes. Pneumonia mortality rates can be reduced with treatments and materials originating from natural resources, such as soy, that can be found in or introduced more broadly across the Nigerian landscape. The use of phospholipid liposomes from soybean byproducts could transform clinical outcomes and reduce mortality rates in children by treating pneumonia with a single dose of antibiotics (Meers et al. 2008). The application of PLA sutures for wound closure and healing can serve as an excellent option compared to traditional sutures. PLA sutures display an excellent biocompatibility complex and can effectively reduce the incidence of acute inflammation and improve wound outcomes. With traumatic injuries, time is of the essence, and deployment of proper resources at primary healthcare facilities outfitted with low-cost and accessible bio-based medical

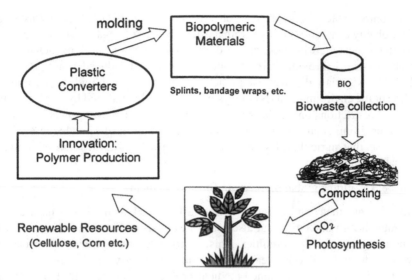

Fig. 4.1 Model for increased sustainability with materials engineered from renewable resources (Mohanty et al. 2002)

supplies and devices can effectively reduce mortality rates and maintain these institutions' operating capacities. The implementation of novel, frugally engineered biomaterial diagnostic devices to detect diarrheal illness could properly identify causal pathogens, improving treatment delivery and outcomes for patients (Bhatia 2010).

- Goal 5: Improve maternal health

Healthcare facilities and homes in Nigeria can be unsterile and in some cases even serve as a source for deadly infections (Keusch et al. 2006). Cleaner conditions for delivery in any setting could drastically improve maternal health for mothers giving birth. Polymer-based biomaterials can be utilized in personal protective equipment (PPE) such as medical gloves, which can be easily recycled after use and limit bodily fluid exchange that could lead to material infection (Shen et al. 2009). In addition, Bhatia and Ramadurai discussed the deployment of 3D printed PLA umbilical cord clamps that are extremely cheap and high in functional utility. These umbilical cord clamps can be fabricated for less than 10 cents and can be utilized by healthcare workers to effectively cut the placenta and free the baby from the womb of the mother. These clamps are biocompatible, display high mechanical strength, and are compostable, thus limiting medical device waste.

- Goal 7: Ensure environmental sustainability

The advantage of using bio-based materials, such as soybeans, corn, kenaf, and silk as starting materials over chemical synthesis techniques to produce products with the same chemical makeup, is that much of the work for the desired product has already been done naturally and without the use of harsh synthesizing agents (Reddy et al. 2012; Weber 2000). The biomaterials we have examined in this book can be compostable or recycled to generate new products (Fig. 4.1) (Vink et al. 2003; Mohanty et al. 2002). This eliminates the concern of negative secondary environmental effects, and in some cases, the recycled material may in fact improve current conditions through more efficient filtration and reduced runoff (Ekebafe et al. 2011). Nigeria will have to approach agricultural growth with a sharp understanding of the environmental risks if systems are poorly managed by implementing policy and restrictions on fertilizer use (Ogburn 2010). The most powerful tool for ensuring environmental sustainability may be the sharing of knowledge and best practices so that agricultural skills and expertise are widespread in community settings (Juma 2010).

- Goal 8: Develop a global partnership for development

Not only are many individuals in Nigeria burdened with deadly diseases and undesirable health outcomes, but also many health specialists leave the country to pursue more lucrative opportunities. This reality, which we previously discussed, known as "brain drain" often leaves individuals in poverty-stricken areas without adequate access to properly trained medical professionals to tend to their ailments (Beine et al. 2001). Biomaterials science can transform this outcome into what we refer to as a "brain gain" by making the necessary resources to provide effective care readily available so that medical professional trainings in Nigeria are incentivized to stay and put their expertise to practice. A few Nigerian Universities, the University of Ibadan and Lagos University, are leading the way in this goal by partnering with engineering and business programs at Northwestern University. The aim of this initiative is to help Nigerian Universities establish biomedical engineering curriculums and facilitate communication between Nigerian engineers, scientists, and clinicians (Paul 2013). Nigerian Universities are full of talented, knowledgeable, and well-trained students and professionals. Where the partnership has been helpful is by introducing connections between Nigerian engineering schools and medical schools. These collaborations have allowed Nigerian researchers and professionals to experience and learn a model for biomedical innovation that they can adapt to their own needs (Glucksberg 2015). The ability to act as a resource and to teach rather than tell is a crucial component that should be considered with global partnerships between developed and emerging economies.

Biomaterials science quite literally gives Nigeria the opportunity to refocus and possibly restructure parts of their healthcare system and treatment strategies from the ground up. By taking advantage of the cultivable land that surrounds them and thinking about agriculture in a new way, they can begin to see, understand, and revolutionize how bio-based materials can be used to create positive outcomes. It must be noted that the materials, concepts, and applications described in this case study are solely additional resources for Nigeria to consider. Healthcare is a complex sector to address, and the adoption of biomaterials into certain parts of the system could reduce material dependency and increase self-reliance for many emerging countries looking for novel ways to address the functional burdens of disease. As authors of this book, we are excited to see the future trajectory that Nigeria will take in securing the health of its population and truly hope that the ideas, concepts, and innovations that we have explored can serve as an impetus for intellectual inquiry and exchange.

Glossary

African Union Promoting unity and solidarity among African states in addition to international cooperation within the framework of the United Nations.

Alveoli and Alveolar Small spaces of the lung where gas exchange occurs.

Amphiphilic Molecules consisting of a water-soluble end and a water-insoluble end.

ANDi operation African Network for Drugs and Diagnostic Innovation (ANDi), an organization promoting pharmaceutical innovation through discovery, development, and delivery of affordable new tools.

Anionic Negatively charged ions.

Biocides and Herbicides Chemical substances that are used to control against harmful organisms. Commonly used in agriculture to deter pests.

Biocompatible The ability of a material to be accepted by the host biological environment, without any adverse reactions.

Biodegrading Materials that when left alone will be broken down by natural biological processes.

Biofilm A dense cluster of cells surrounding an infected area.

Biomaterials and Bio-based Materials Engineered materials that are derived in whole, or in part, from renewable sources and living mater. These two terms will be used interchangeably throughout this case study.

Biomaterials Science The study of molecular, chemical, and physical interactions of biomaterials and their many applications.

Brain Drain The migration of educated and high impact professionals from developing areas due to limited opportunities.

Brain Gain The idea that with greater opportunities and resources, high impact professionals will remain in their native countries.

Duodenum and Distal Colon The duodenum is the first and shortest section of the small intestine and plays a vital role in chemical digestion. The distal colon is the last part of the colon and large intestine and is vital for nutrient and water absorption.

© The Author(s) 2018   47
A.A. Tracy et al., *Bio-Based Materials as Applicable, Accessible, and Affordable Healthcare Solutions*, SpringerBriefs in Materials,
https://doi.org/10.1007/978-3-319-69326-2

ESSENCE Enhancing Support for Strengthening the Effectiveness of National Capacity Efforts. A TDR-based initiative aligning research programs with the priorities of disease-endemic countries.

Etiologic Microbes Causal organisms and the origin of diseases.

European Union The political and economic union of member states located in Europe.

Greenhouse Gas A gas that absorbs and emits radiation that is within the thermal infrared range.

Hydrophilic Having a strong affinity to water.

Imaging agents Chemicals or molecules designed to target specific tissue types for diagnosis and monitoring health conditions with imaging instruments.

ISHReCA Initiative to Strengthen Health Research Capacity in Africa. This organization is focused on improving communication between research and funding for health research.

Kenaf A nonedible plant similar to bamboo.

Microfluidics A multidisciplinary field of science that involves the use of small volumes, small size, and low-energy consumption.

Mucinous material A fluid buildup surrounding an area of infection.

Pathogen An infectious biological agent that causes disease and illness.

Polymeric matrix Dense and hard polymer structures secreted by bacterial infections.

Respiratory Distress Syndrome Commonly occurs in infants whose lungs have not fully developed where the lack of surfactant causes trouble in filling and maintaining air in the lungs.

Reverse Osmosis A water purification technology that uses a membrane to remove particles.

SC-CO$_2$ Supercritical carbon dioxide is a fluid state of carbon dioxide held at very high temperature and pressure. It is used as an industrial solvent for chemical extraction processes.

Smart Plastics Industry-changing nanotechnology applications that address the growth of global water and energy demand made by the Dias Analytic Corporation

Smart technology Technologies with advanced analytical software and computational power.

Successful technology Spin-off of "smart technology," this technology is simplified to achieve the basic task at hand with minimal input, resources, and manuals.

Systemic Septicemia A severe blood infection that can lead to organ failure.

TDR The Special Program for Research and Training in Tropical Diseases that helps facilitate and support research efforts to reduce the burden of diseases caused by poverty.

UNCPD The United Nations Commission on Population and Development is a group that investigates population issues and trends.

UNDP The United Nations Development Program. This organization helps build nations to deal with crisis and improve the quality of life.

UNECA United Nations Economic Commission for Africa. This council promotes economic and social development for African UN member states.

Water and Sanitation Program A multi-donor partnership that is part of the World Bank that supports and promotes water and sanitation services.

References

Aall, C. (1970). Relief, nutrition and health problems in the Nigerian/Biafran war. *Journal of Tropical Pediatrics, 16*(2), 70–90.

African Network for Drugs and Diagnostic Innovation. (2010). www.andi-africa.org

African Progress Panel. (2014). *Grain fish money. Financing Africa's green and blue revolutions.* Africa Progress Report 2014.

Akinpelu, A. O., Alonge, T. O., Adekanla, B. A., & Odole, A. C. (2009). Prevalence and pattern of symptomatic knee osteoarthritis in Nigeria: A community-based study. *The Internet Journal of Allied Health Sciences and Practice, 7*(3), 1–7.

Athanasiou, K. A., Agrawal, C. M., Barber, F. A., & Burkhart, S. S. (1998). Orthopaedic applications for PLA-PGA biodegradable polymers. *Arthroscopy: The Journal of Arthroscopic & Related Surgery, 14*(7), 726–737.

Auras, R. A., Lim, L. T., Selke, S. E., & Tsuji, H. (Eds.). (2011). *Poly (lactic acid): Synthesis, structures, properties, processing, and applications* (Vol. 10). Hoboken: Wiley.

Avella, M., Bogoeva-Gaceva, G., Bužarovska, A., Errico, M. E., Gentile, G., & Grozdanov, A. (2008). Poly (lactic acid)-based biocomposites reinforced with kenaf fibers. *Journal of Applied Polymer Science, 108*(6), 3542–3551.

Bakker-Woudenberg, I. A. J. M., Lokerse, A. F., Ten Kate, M. T., Mouton, J. W., Woodle, M. C., & Storm, G. (1993). Liposomes with prolonged blood circulation and selective localization in Klebsiella pneumoniae-infected lung tissue. *Journal of Infectious Diseases, 168*(1), 164–171.

BBC News Africa. (2013). www.bbc.co.uk/news/world-africa-13951696

Beine, M., Docquier, F., & Rapoport, H. (2001). Brain drain and economic growth: Theory and evidence. *Journal of Development Economics, 64*(1), 275–289.

Bhatia, S. K. (2010). *Biomaterials for clinical applications.* New York: Springer.

Bhatia, S. K. (2012). *Bio-based materials step into the operating room.* American Institute of Chemical Engineers. CEP. (pp. 49–53).

Bhatia, S. K., & Ramadurai, K. W. (2017). 3-Dimensional device fabrication: A bio-based materials approach. In *3D printing and bio-based materials in global health* (pp. 39–61). Cham: Springer International Publishing.

Bhatia, S. K., Arthur, S. D., Chenault, H. K., & Kodokian, G. K. (2007). Interactions of polysaccharide-based tissue adhesives with clinically relevant fibroblast and macrophage cell lines. *Biotechnology Letters, 29*(11), 1645–1649.

Bhatti, Y. A. (2012). What is frugal, what is innovation? Towards a theory of frugal innovation. Oxford Centre for Entrepreneurship and Innovation. (Unpublished).

© The Author(s) 2018
A.A. Tracy et al., *Bio-Based Materials as Applicable, Accessible, and Affordable Healthcare Solutions,* SpringerBriefs in Materials,
https://doi.org/10.1007/978-3-319-69326-2

Bhatia, S. K., & Ramadurai, K. (2017). *3D printing and bio-based materials in global health.* Springer Briefs in Materials.

Black, R. E., Morris, S. S., & Bryce, J. (2003). Where and why are 10 million children dying every year? *The Lancet, 361*(9376), 2226–2234.

Bloom, D. E., & Williamson, J. G. (1998). Demographic transitions and economic miracles in emerging Asia. *The World Bank Economic Review, 12*(3), 419–455.

Bodeker, G., Ryan, T., & Ong, C. K. (1999). Traditional approaches to wound healing. *Clinics in Dermatology, 17*(1), 93–98.

Bugge, M. M., Hansen, T., & Klitkou, A. (2016). What is the bioeconomy? A review of the literature. *Sustainability,* 8(7), 691.

Costerton, J. W., Stewart, P. S., & Greenberg, E. P. (1999). Bacterial biofilms: A common cause of persistent infections. *Science, 284*(5418), 1318–1322.

Christensen, C., & Raynor, M. (2013). The Innovator's solution: Creating and sustaining successful growth. Boston: Harvard Business Press.

Dada, E., Erinne, J., & Taiwo, O. (2013). Chemical engineering in Nigeria: Development, challenges, and prospects. *American Institute of Chemical Engineers, 109,* 52–56.

Dare, A. J. (2015). *Lancet commission on global surgery economics & financing.* Retrieved Dec 02 2016, from http://www.globalsurgery.info/wp-content/uploads/2015/02/Session-14_Economics-Financing-presentation.pdf

Dasaraju, P. V., & Liu, C. (1996). Infections of the respiratory system. In S. Baron (Ed.), *Medical microbiology* (4th ed.). Galveston: University of Texas Medical Branch.

Edmond, K., Scott, S., Korczak, V., Ward, C., Sanderson, C., Theodoratou, E., et al. (2012). Long term sequelae from childhood pneumonia; systematic review and meta-analysis. *PLoS One, 7*(2), e31239.

Ekebafe, L. O., Ogbeifun, D. E., & Okieimen, F. E. (2011). Polymer applications in agriculture. *Biokemistri, 23*(2), 81–89.

Evans, D., Edejer, T., Chisholm, D., & Stanciole, A. (2009). *WHO guide to identifying the economic consequences of disease and injury.* Geneva: Department of Health Systems Financing, Health Systems and Services, World Health Organization.

Fatunde, O. A., & Bhatia, S. K. (2012). *Medical devices and biomaterials for the developing world: Case studies in Ghana and Nicaragua.* New York: Springer Briefs in Public Health.

Fraunhofer Life Sciences Engineering Group. (2008). http://www.fhcmi.org/LSE/Projects/08.html

George, R. (2009). The politics of toilets. *The Washington Post. Post Global.*

Greenwood, H. L., Singer, P. A., Downey, G. P., Martin, D. K., Thorstelnsdottlr, H., & Daar, A. S. (2006). Regenerative medicine and the developing world. *PLoS Medicine, 3*(9), e381.

Glucksberg, M. (2015). Center for Innovation in Global Health Technologies. Northwestern McCormick School of Engineering and Applied Science. Personal Phone Interview. Thursday, February 12th.

Guerrant, R. L., Oriá, R. B., Moore, S. R., Oriá, M. O., & Am Lima, A. (2008). Malnutrition as an enteric infectious disease with long-term effects on child development. *Nutrition Reviews, 66*(9), 487–505.

Helen, C. M. (Ed.). (1991). *Nigeria: A country study.* Washington: GPO for the Library of Congress.

Hofman, K., Primack, A., Keusch, G., & Hrynkow, S. (2005). Addressing the growing burden of trauma and injury in low-and middle-income countries. *American Journal of Public Health, 95*(1), 13.

Institute for Economics and Peace. (2012). *Global Terrorism Index—Capturing the Impact of Terrorism from 2002–2011.* www.visionofhumanity.org

International Institute of Tropical Agriculture (IITA). (2009a). www.iita.org/maize

International Institute of Tropical Agriculture (IITA). (2009b). www.iita.org/soybean

Jones, M. N., Song, Y. H., Kaszuba, M., & Reborias, M. D. (1997). The interaction of phospholipid liposomes with bacteria and their use in the delivery of bactericides. *Journal of Drug Targeting, 5*(1), 25–34.

Juma, C. (2010). *The new harvest: agricultural innovation in Africa.* New York: Oxford University Press.

Kenaf fiber-reinforced polylactic acid: Research & Development I NEC. (n.d.). Retrieved 08 Sept 2017, from http://www.nec.com/en/global/rd/innovation/bioplastics/page03_1.html

Keusch, G. T. (2001). *Toxin-associated gastro-intestinal disease: A clinical overview. Molecular medical microbiology*. New York: Academic Press.

Keusch, G. T., Fontaine, O., Bhargava, A., Boschi-Pinto, C., Bhutta, Z. A., Gotuzzo, E., et al. (2006). Diarrheal diseases. In D. T. Jamison, J. G. Breman, A. R. Measham, et al. (Eds.), *Disease control priorities in developing countries* (2nd ed.). Washington: World Bank.

Labinjo, M., Juillard, C., Kobusingye, O. C., & Hyder, A. A. (2009). The burden of road traffic injuries in Nigeria: Results of a population-based survey. *Injury Prevention, 15*(3), 157–162.

Lasprilla, A. J. R., Martinez, G. A. R., & Hoss, B. (2011). Synthesis and characterization of poly (lactic acid) for use in biomedical field. *Chemical Engineer, 24*, 985–990.

Leke, A., Fiorini, R., Dobbs, R., Thompson, F., Suleiman, A., & Wright, D. (2014). *Nigeria's renewal: Delivering inclusive growth in Africa's largest economy*. McKinsey Global Institute. www.mckinsey.com/global-themes/middleeast-and-africa/nigerias-renewal-delivering-inclusive-growth.

Li, J., He, Y., & Inoue, Y. (2003). Thermal and mechanical properties of biodegradable blends of poly (L-lactic acid) and lignin. *Polymer International, 52*(6), 949–955.

List, G. R., & Friedrich, J. P. (1989). Oxidative stability of seed oils extracted with supercritical carbon dioxide. *Journal of the American Oil Chemists Society, 66*(1), 98–101.

Maric, J., Rodhain, F., & Barlette, Y. (2016). Frugal innovations and 3D printing: Insights from the field. *Journal of Innovation Economics & Management, 3*, 57–76.

Marshall, J. A. (2002). Mixed infections of intestinal viruses and bacteria in humans. In K. A. Sussman Brogden & J. M. Guthmiller (Eds.), *Polymicrobial diseases*. Washington: ASM Press.

Matthews, P., Ryan-Collins, L., Wells, J., Sillem, H., & Wright, H. (2012). *Engineers for Africa: Identifying engineering capacity needs in sub-Saharan Africa. A Summary Report*. London: Africa-UK Engineering for Development Partnership/Royal Academy of Engineering.

Meers, P., Neville, M., Malinin, V., Scotto, A. W., Sardaryan, G., Kurumunda, R., et al. (2008). Biofilm penetration, triggered release and in vivo activity of inhaled liposomal amikacin in chronic Pseudomonas aeruginosa lung infections. *Journal of Antimicrobial Chemotherapy, 61*(4), 859–868.

Merolli, A., Nicolais, L., Ambrosio, L., & Santin, M. (2010). A degradable soybean-based biomaterial used effectively as a bone filler in vivo in a rabbit. *Biomedical Materials, 5*(1), 015008.

Monroe, D. (2007). Looking for chinks in the armor of bacterial biofilms. *PLoS Biology, 5*(11), e307.

Montanari, L., King, J. K., List, G. R., & Rennick, K. A. (1996). Selective extraction of phospholipid mixtures by supercritical carbon dioxide and Cosolvents. *Journal of Food Science, 61*(6), 1230–1234.

Mohanty, A. K., Misra, M., & Drzal, L. T. (2002). Sustainable bio-composites from renewable resources: Opportunities and challenges in the Green Materials World. *Journal of Polymers and the Environment., 10*, 19.

Mulholland, E. K., Simoes, E. A. F., Costales, M. O. D., McGrath, E. J., Manalac, E. M., & Gove, S. (1992). Standardized diagnosis of pneumonia in developing countries. *The Pediatric Infectious Disease Journal, 11*(2), 77–81.

Ochi, S. (2008). Mechanical properties of kenaf fibers and kenaf/PLA composites. *Mechanics of Materials, 40*(4), 446–452.

Ogbomo, S. M., Ayre, B., Webber, C. L., & D'Souza, N. A. (2014). Effect of kenaf fiber age on PLLA composite properties. *Polymer Composites, 35*(5), 915–924.

Ogburn, S. (2010). *Fertile ground: The dark side of nitrogen*. Grist. http://grist.org/article/2009-11-11-the-dark-side-of-nitrogen/

Olomola, A. S. (2007). Strategies for managing the opportunities and challenges of the current agricultural commodity booms in SSA. *Seminar papers on managing commodity booms in sub-Saharan Africa*. A Publication of the AERC Senior Policy Seminar IX, African Economic Research Consortium (AERC), Nairobi.

Orekunrin, O. (2013, October 18). Africa's trauma epidemic. The Opinion Pages. A27. *The New York Times.*

Paul, M. (2013). Northwestern helps bring first biomedical engineering programs to Nigerian universities. *McCormick News.* Northwestern Engineering. http://www.mccormick.northwestern.edu/news/articles/2013/11/northwestern-helps-bring-first-biomedical-engineering-programs-to-nigerian-universities.html

Perut, F., Montufar, E. B., Ciapetti, G., Santin, M., Salvage, J., Traykova, T., et al. (2011). Novel soybean/gelatine-based bioactive and injectable hydroxyapatite foam: Material properties and cell response. *Acta Biomaterialia, 7*(4), 1780–1787.

Rafindadi, A. H. (2004). A review of types of injuries sustained following road traffic accidents and their prevention. *Nigerian Journal of Surgical Research, 2*(3), 100–104.

Rankin, T. M., Giovinco, N. A., Cucher, D. J., Watts, G., Hurwitz, B., & Armstrong, D. G. (2014). Three-dimensional printing surgical instruments: Are we there yet? *Journal of Surgical Research, 189*(2), 193–197.

Reddy, M. M., Misra, M., & Mohanty, A. K. (2012). Bio-based materials in the new bio-economy. *Chemical Engineering Progress, 108*(5), 37–42.

Rockwood, D. N., Preda, R. C., Yücel, T., Wang, X., Lovett, M. L., & Kaplan, D. L. (2011). Materials fabrication from Bombyx Mori silk fibroin. *Nature Protocols, 6*(10), 1612–1631.

Sanadi, A. R., Caulfield, D. F., Jacobson, R. E., & Rowell, R. M. (1995). Renewable agricultural fibers as reinforcing fillers in plastics: Mechanical properties of kenaf fiber-polypropylene composites. *Industrial & Engineering Chemistry Research, 34*(5), 1889–1896.

Santin, M., & Ambrosio, L. (2008). Soybean-based biomaterials: Preparation, properties and tissue regeneration potential. *Expert Reviews in Medical Devices, 5*(3), 349–358.

Santin, M., Morris, C., Standen, G., Nicolais, L., & Ambrosio, L. (2007). A new class of bioactive and biodegradable soybean-based bone fillers. *Biomacromolecules, 8*(9), 2706–2711.

Shen, L., Haufe, J., & Patel, M. K. (2009). *Product overview and market projection of emerging bio-based plastics.* Utrecht: European Bioplastics and European Polysaccharide Network of Excellence.

Shih, Y. F., & Huang, C. C. (2011). Polylactic acid (PLA)/banana fiber (BF) biodegradable green composites. *Journal of Polymer Research, 18*(6), 2335–2340.

Sittoni, T., & Maina, S. (2012). *Nigeria loses NGN455 billion annually due to poor sanitation. Economic impacts of poor sanitation in Africa.* Washington: World Bank.

Smith, A. W. (2005). Biofilms and antibiotic therapy: Is there a role for combating bacterial resistance by the use of novel drug delivery systems? *Advanced Drug Delivery Reviews, 57*(10), 1539–1550.

Speizer, F. E., Horton, S., Batt, J., & Slutsky, A. S. (2006). Respiratory diseases of adults. In D. T. Jamison, J. G. Breman, A. R. Measham, et al. (Eds.), *Disease control priorities in developing countries* (2nd ed.). Washington: World Bank.

Stewart, P. S., & Costerton, J. W. (2001). Antibiotic resistance of bacteria in biofilms. *The Lancet, 358*(9276), 135–138.

Stock, R. (1983). Distance and the utilization of health facilities in rural Nigeria. *Social Science & Medicine, 17*(9), 563–570.

Surace, R., Trotta, G., Bellantone, V., & Fassi, I. (2012). The micro injection moulding process for polymeric components manufacturing. In C. Volosencu (Ed.), *New technologies—Trends, innovations and research.* Manhattan: InTech.

Sutherland, I. W. (2001). The biofilm matrix—an immobilized but dynamic microbial environment. *Trends in Microbiology, 9*(5), 222–227.

TEDxBoston. (2012). www.ted.com/talks/timothy_prestero_design_for_people_not_awards.html

Toner, M., & Irimia, D. (2005). Blood-on-a-chip. *Annual Review of Biomedical Engineering, 7*, 77.

U.S. Dept. of Agriculture. (2008). *U.S. Biobased products market potential and projections through 2025.* Washington: USDA.

United Nations. (2015). We can end poverty. Millennium development goals and beyond 2015. http://www.un.org/millenniumgoals/

United Nations Development Program (UNDP). (2006). *Human development report 2006: Beyond scarcity, poverty, power and the global water crisis*. New York: UNDP.

Vink, E. T., Rabago, K. R., Glassner, D. A., & Gruber, P. R. (2003). Applications of life cycle assessment to NatureWorks™ polylactide (PLA) production. *Polymer Degradation and Stability, 80*(3), 403–419.

Vré, R. D., Verde, E. R, & da Silva, J. S. (2010). *Closing the R&D gap in African health care*. McKinsey Quarterly. McKinsey Company. www.mckinsey.com/industries/healthcare-systems-and-services/our-insights/closing-the-r-and-38d-gap-in-african-health-care.

Walker, A. (2012). *What is Boko Haram?* Washington: US Institute of Peace.

Webber, C. L. (1999). Effect of kenaf and soybean rotations on yield components. In J. Janick (Ed.), *Perspectives on new crops and new uses* (pp. 316–321). Alexandria: ASHS Press.

Webber, C. L., & Bledsoe, V. K. (2002). Kenaf yield components and plant composition. In J. Janick & A. Whipkey (Eds.), *Trends in new crops and new uses*. Alexandria: ASHS Press.

Weber, C. J. (2000). *Biobased packaging materials for the food industry: Status and perspectives*. Frederiksberg: European Union Directorate 12, Royal Veterinary and Agricultural Univ..

Weigl, B. H., Gerdes, J., Tarr, P., Yager, P., Dillman, L., Peck, R., et al. (2006). Fully integrated multiplexed lab-on-acard assay for enteric pathogens. In *Proc. SPIE 6112, microfluidics, bioMEMS, and medical microsystems IV*, 611202 (23 January 2006); http://dx.doi.org/10.1117/12.644714.

Williams, D. F. (1987). *Definitions in biomaterials: proceedings of a consensus conference of the European Society for Biomaterials* (Vol. 4). Chester: Elsevier Science Ltd.

Woolf, A. D., & Pfleger, B. (2003). Burden of major musculoskeletal conditions. *Bulletin of the World Health Organization, 81*(9), 646–656.

World Bank. (2014). Tackling the road safety crisis in Africa. http://www.worldbank.org/en/news/feature/2014/06/06/tackling-the-road-safety-crisis-in-africa

World Health Organization. (1979). *Formulating strategies for health for all by the year 2000*. Geneva: WHO Press.

World Health Organization. (2006). *Country health system fact sheet 2006—Nigeria*. Geneva: WHO Press.

World Health Organization. (2008). *The global burden of disease: 2004 update*. Geneva: WHO Press.

World Health Organization. (2010a). *Barriers to innovation in the field of medical devices*. Geneva: WHO Press.

World Health Organization. (2010b). *Building bridges for innovation across Africa*. TDR*news*. No.86. Geneva: WHO Press.

World Health Organization. (2012). *Nigeria launches 'saving one million lives' by 2015 initiative*. http://www.who.int/workforcealliance/media/news/2012/1mlives/en/

World Health Organization. (2013a). *Nigeria. Country statistics*.

World Health Organization. (2013b). *Nigeria. Neonatal and child health profile*. www.who.int/entity/maternal_child_adolescent/epidemiology/profiles/neonatalchild/nga.pdf

World Health Organization. (2013c). *Road safety. Estimated number of road traffic deaths, 2010*. www.gamapserver.who.int/gho/interactive_charts/road_safety/road_traffic_deaths/atlas.html

World Health Organization. (2013d). *Special programme for research and training in tropical diseases (TDR)*. www.who.int/tdr/about/en/

World Health Organization. (2014). *Nigeria: Health profile*. http://www.who.int/gho/countries/nga.pdf

Xiao, L., Wang, B., Yang, G., & Gauthier, M. (2012). Poly (lactic acid)-based biomaterials: synthesis, modification and applications. In D. N. Ghista (Ed.), *Biomedical science, engineering and technology*. Rijeka: InTech.

Yager, P., Edwards, T., Fu, E., Helton, K., Nelson, K., Tam, M. R., & Weigl, B. H. (2006). Microfluidic diagnostic technologies for global public health. *Nature, 442*(7101), 412–418.

Young, A. L. (2003). Biotechnology for food, energy, and industrial products new opportunities for bio-based products. *Environmental Science and Pollution Research, 10*(5), 273–276.

Printed in the United States
By Bookmasters